MODELING REALITY

George Dlouhý

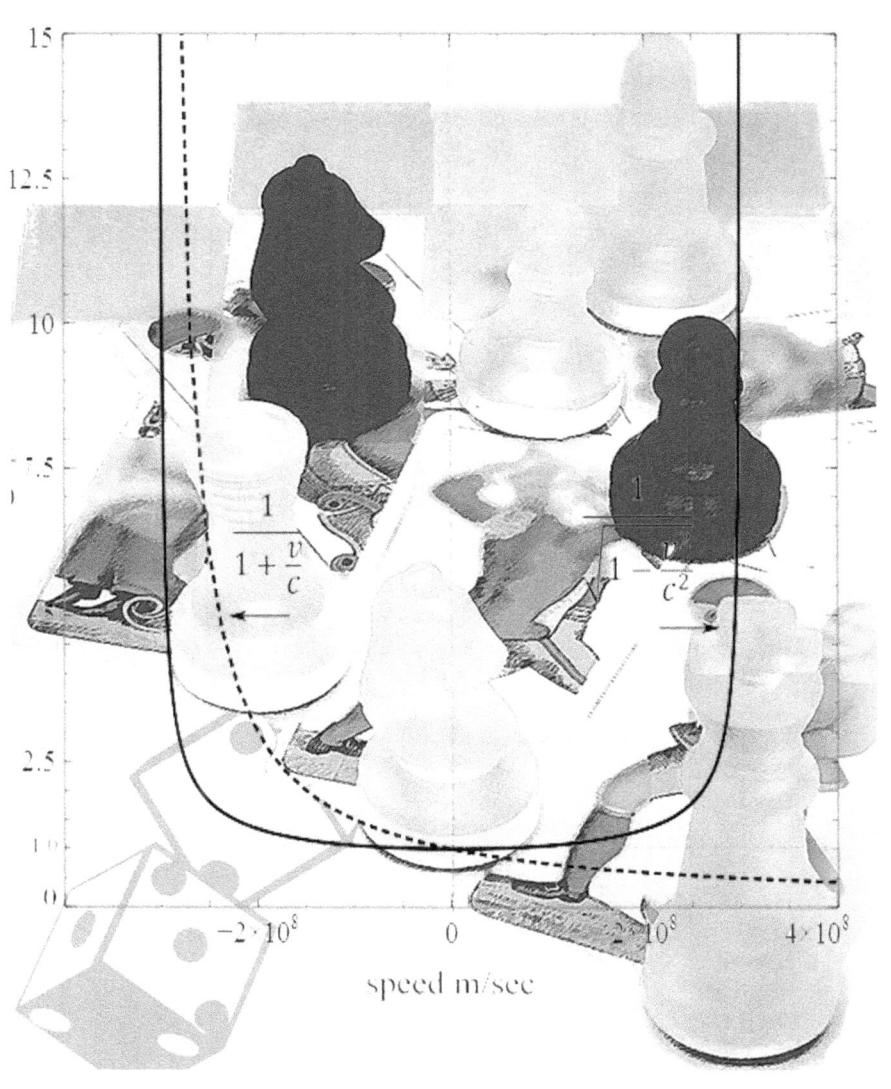

"Imagination is more important than knowledge"
Albert Einstein

© George Dlouhý 2013
www.dlouhy.info

ISBN 978-1484990926
Catalogued by National Library of Australia
First edition July 2013
Seventh revised edition December 2022

Contents

Foreword

The peak of Socrates' wisdom is his famous quote:

"I know that I know nothing."

I believe it was the sheer admiration of complexity of our world and not an impulse of disappointment or sadness that inspired him to say it. And it was this inherited admiration, which created science and developed the human race.

Socratic "nothing" is not physically represented in our world and whatever we undertake, this "nothing" will forever represent our knowledge about the existence of this world.

That does not mean that we should, not at least, try to decipher some of the mysteries enshrouding our three-dimensional world.

This brochure uses simple modeling and tries to simplify many complex, puzzling issues and offers appropriate explanations. It could be broadly compared to a tool, to which we could add a never-ending list of useful attachments. As time and our understanding progress, the ideas in this brochure could be constantly enhanced, perfected and possibly some dismissed. The content of this brochure is only a tiny foundation, an open code that could be further modified and developed.

There is nothing new about modeling reality. It has been put to practice since humans first invented toys. For children toys mimic reality and possess many characteristics taken from the real world. On these simplified models, on replicas of some real, complex objects and situations, children learn to live with the realities of our world. The modeling in this brochure follows a similar scenario, analyzing real objects and situations by using abstract and real models. With their help the brochure ambitiously takes on revealing, or at least simplifying, some aspects of our existence on this planet.

The modeling used in this brochure derives tangible guidelines, which lead to the conclusion that our world is just a computer simulation or just a mere game, being played in another world.

Computers and their programming started many decades ago and I am well aware that I am not the first to suggest a comparison between our world and the digital world. There must be many of us puzzled by such close resemblance, especially when towards the end of the twentieth century the new paradigm, called Object Oriented Programming, was introduced. That made the correlation between modeling of objects existing in our world and objects in the digital world of computers more distinct and tempting to draw parallels between them.

The first edition of this brochure was meant to contain just guidelines, offered to anybody to elaborate and create a qualified discussion. As time progressed, I realized that I was asking too much; this brochure did not raise expected discussion and some of my derived conclusions were even wrong.

As a result of that, I naively believed that I should enhance the digital model of our world by adding some necessary calculations and graphs. Obviously, this was a mistake and I should have adhered to the generally accepted belief that *"Even one equation will render any book unpopular"*.

This seventh edition of this brochure is therefore the first combining this new concept into a series of easily understandable steps and is based mainly on logical reasoning.

1. Model Used

If we wish that our modeling of some given situation would follow a realistic path as close as possible, then our model has to follow the same path. We cannot, for example, successfully create a model of inter-human relations based on animals' behavior in a zoological garden.

We have to establish that some basic rules applicable to our world also apply to our model. For a greater degree of accuracy, our model should contain many existing attributes of our world, but for the sake of simplicity our model should contain their smallest number possible. We could then leave out irrelevant attributes and concentrate only on basic, desirable attributes, like for example numbers. Since Galileo Galilei already believed that *Mathematics is the language with which God has written the universe*, we should select simple numbers as the building block of our model.

One example of such simplified modeling of numbers is an abstract numerical line where every number is represented by a distance from a point, representing the number zero. Such a line is only one-dimensional and could not exist in our three-dimensional world. In our world it could be represented, for example, by a school ruler, containing a small subset of numbers. We could find numbers 0,1, 2, etc. and the ruler usually ends with number 30.

Most rulers also have fractions, like 1.5 or 2.1. So far, all these numbers could be expressed as a fraction of two whole numbers, i.e., 15/10 or 21/10. Long time ago it was decided that such numbers will be called *rational numbers*, since in the opinion of mathematicians, these numbers behave rationally.

There exist also *irrational numbers*, which could not be expressed as a fraction of two whole numbers and their number of decimal places never ends. One example of such a number is the square root of two √2.

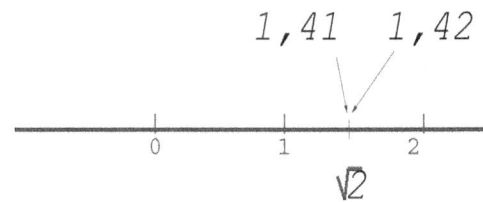

Fig. 1.1 *Numeric line and number √2, situated between 1,41 and 1,42*

Most electronic calculators would display this number as 1.4142, but if you have a more advanced calculator, this number could be 1.414213562. Although the number of decimal places of this number never ends, this number still should exist on a numerical line and should be represented by the length between 0 and √2. This number is situated somewhere between 1.4142 and 1.4143, more precisely between 1.414213 and 1.414214, and even more precisely, between 1,414213562 and 1,414213563. In this manner we could continue indefinitely and therefore this number exists only in infinity.

This raises a question: How could infinity exist on a one-dimensional number line? Since there was not any plausible explanation, mathematicians previously agreed that irrational numbers will not be defined on the numerical line.

Yet, in our mind we could, for example, multiply two square roots of two ($\sqrt{2} \times \sqrt{2} = 2$) and therefore our mind must also contain irrational numbers. That would imply that our mind must be infinite and the numerical line containing the irrational numbers must be also infinite.

In our world we could not create an infinite line, but we could create its three-dimensional representation, i.e., on a blackboard we could draw a line. We could also represent some functions, like for example the multiplication of two numbers as a defined size of an area on the blackboard. Then the *observers* of that blackboard will create this line and area in their mind and the associated numbers would be then available in our three-dimensional world.

We could be satisfied that our two-dimensional model simplifies the concept of numbers and it represents them satisfactorily. Since our selected model should explain more than just this concept, we would have to create a more complex model than the numeric line.

Fortunately, a revolutionary new concept has been developed, which changed the very basic values ruling our civilization, namely "yes" and "no", into mathematical numbers 1 and 0. I believe that was the very first time when existing values in our world were transferred into a model where they could be represented and manipulated. We could not, for example, add, subtract, multiply or divide "yes" and "no", unless we associate them in our mind with numbers, as is done by the computer.

This model, having only two values, was called digital and since its introduction it serves as a base for what is now generally referred to as digital computing.

There is a groundbreaking difference between our world of "yes" and "no", and this digital model. When the digital model expresses "yes" and "no" as series of 1 and 0 (called digital numbers), we could perform mathematical functions on processed data. We could, for example, represent the weight of one particular horse and of one particular elephant by a row of numbers 1 and 0, corresponding to the number of units of weight. By using addition and division we could calculate that the elephant is bigger than the horse. This statement is based on our model and is supported by simple mathematical operations. This is of course, not possible in our world of "yes" and "no".

We now have a digital model, which could tell us more about our world than we could deduce by simple observations, and we could use this digital model as a base for our modeling. Of course, the modeling of our world based on this simple model will be inadequate for the purpose of this book, since it will give us only a very basic understanding.

Fortunately, since its introduction, the digital computing has rapidly developed into a perfect tool for our modeling. All we need is an imaginary computer, based on digital computing, equipped with an imaginary programming language.

This enables us to understand the processes behind the existence of some unexplained phenomena, which we still encounter and so far, have not found any solid, logical explanation for their existence.

2. Our Mind Is Infinite

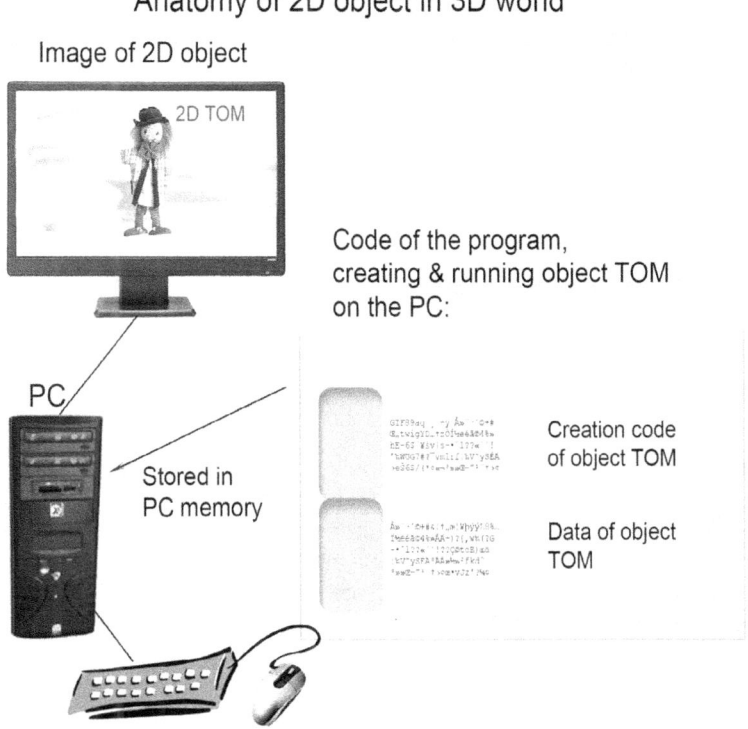

Fig. 2.1 Computer-generated object Tom, existing in the 3D world consists of::
1. Tom's 2D image displayed on the screen.
2. Instruction set containing code needed to create object Tom.
3. Data associated with the object, i.e., object's properties.

There is one limitation to our model, though. Since the objects in our programming model could be displayed in our world on a computer monitor as two-dimensional images only, our model is therefore only two-dimensional (2D), not three dimensional (3D), as our world. This limitation is actually a welcomed feature, since it brings substantial simplification to our modeling.

In the initial stages, this new digital computing was hindered by the simplicity of tools available. Only towards the end of the last century a new paradigm governing programming, called *Object Oriented Programming,* was developed (*OOP*). This model uses real objects in our three-dimensional world as a blueprint for structuring programs executed on computers.

When such a program is running, an object, for example "factory", could create some other specific objects. These objects consist of a block of code, stored in computer memory and written in programming language, and they could also have their image displayed on the computer terminal.

We are already familiar with pseudo two-dimensional modeling, because such models are all around us: photographs, paintings, films, computer games etc.

Our model with all existing objects is also two-dimensional, contravening our world where all objects are three-dimensional. Therefore in our three-dimensional world these objects cannot exist. They are only represented by their two-dimensional images on the computer screen, which are nothing else but emitted rays of light.

Fig. 2.2 Tom observes Bunny's image existing in a two-dimensional world.

In the figure above, the observed Bunny's image is only two-dimensional and is a part of Bunny's program code, stored in the computer in our 3D world. While Tom observes this image, Bunny's two-dimensional image exists in his mind.

We have already added numbers to our model, and we could even produce their representation on a numeric line, drawn on a blackboard. In reality, this line is not one-dimensional at all. Since it is a part of our three-dimensional world, it could be subjected to all possible physical operations on three-dimensional objects. For example, we could rub off a part of that line.

In figure *1.1* we selected numbers 1 and 2, having their representation displayed on a numeric line. Somewhere in between them lies another number, namely the already mentioned square root of two ($\sqrt{2}$).

Using Pythagoras' theory, this number could be represented by a line, connecting two points of a triangle, as depicted in the following figure:

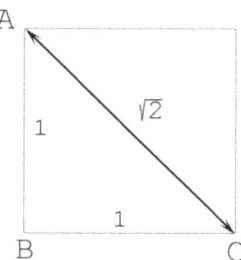

Fig. . 2.3 Distance AB = 1, BC = 1, $\sqrt{2}$ is represented by distance AC and angle ABC is 90°. According to Pythagoras' law, the hypotenuse is equal to $\sqrt{2}$.

$$AC^2 = AB^2 + BC^2$$
$$\textit{since } AB = BC = 1, \textit{ then}$$
$$AC^2 = 1^2 + 1^2 = 2$$
$$AC = \sqrt{2}$$

When we draw a line in our three-dimensional world, we could never be certain about its actual length. If we measure it by a school ruler, we could agree that the line is one unit length long. If we use a magnifying glass and look again at the actual beginning and ending of the line, we will find that this is not the case. Both positions do not exactly match the marks on the ruler.

We could use a more precise ruler and keep on increasing magnification and yet, we will never achieve the exact measure. It is because the number $\sqrt{2}$ is never ending, it's exact value lies in infi*nity* and the number itself is infinite.

Despite that, this number is still represented by some distance on the numerical line, created in our mind.

By sing our model, we have come to the first valid conclusion:

Our mind is infinite and therefore it could not be an integral part
of our three-dimensional world.

3. Universe Is Not Infinite

From now on we accept that our mind is infinite and is not a part of our three-dimensional universe. Should our mind be a part of our universe, then our universe would have to be also infinite.

Fortunately, there is a simple proof for our reasoning so far, confirming that the universe is not infinite:

One of the main attributes of the universe is the light, which travels through the vacuum filling the universe with its constant speed $\underline{c}$.

That could be expressed as the distance D, travelled by the light, divided by time $\underline{t}$ it takes for the light to travel this distance $c = D / t$

Should the universe be infinite, then the light can travel an infinite distance
$$D = \infty$$

For the light to travel an infinite distance, it will take an infinite time $t = \infty$

The speed of light then becomes $c = \infty / \infty$

This ratio of two infinities is not defined, since ∞ / ∞ cannot be equal to 1.

(If it is equal, then the following will apply: $\infty / \infty = 1.$

Since $\infty + \infty = \infty$,

then $\infty / \infty = (\infty + \infty) / \infty = (\infty / \infty) + (\infty / \infty) = 1 + 1 = 2$

The following will also apply: $\infty + \infty + \infty = \infty$

$\infty / \infty = (\infty + \infty + \infty) / \infty = (\infty / \infty) + (\infty / \infty) + (\infty / \infty) = 1 + 1 + 1 = 3)$

This obviously cannot be valid and therefore the speed of light in infinity cannot be defined, i.e., the light does not exist in infinity. Since light exists in our universe, the unavoidable conclusion is that our universe cannot be infinite.

Fig. 3.1 *Our universe is a mere bubble, filled with vacuum and immersed in the infinite world.*

4. Our Brain and Mind

So far, we have proved that our mind exists in infinity and our world is not infinite. Inevitably, our mind does not exist in our world and thus there must be some connection between our body and our mind.

Given that our interaction with our world is done through our sensors, like eyes, ears, touch, etc., which are all connected to our brain, it must be the brain, which communicates with our mind. To analyze this hypothesis, we will resort to our digital model:

Tom, *Mary* and *Bunny* are objects created by the program and are represented in a 2D world by their two-dimensional images on a computer screen. The program code and data of these objects reside in computer memory, i.e., in a 3D world.

Tom observes *Mary* and *Bunny's* images, and if this situation should resemble reality, *Tom's* brain has to be a part of *Tom's* image and resides on the screen in *Tom's* head. Unfortunately, *Tom's* image is only a collection of miniature pixels emitting colored light and having very limited functionality. For our model to be a true representation of the situation in our world, *Tom's* brain, existing on a 2D screen in *Tom's* image, should function similarly to the human brain.

We could hypothetically assume that *Tom's* brain has such capability and could register *Mary's* and *Bunny's* existence and store all the information *Tom* could find about them in the computer memory in the 3D world. To fully comply with the situation in our orld, *Tom* has also to remember and recall this information.

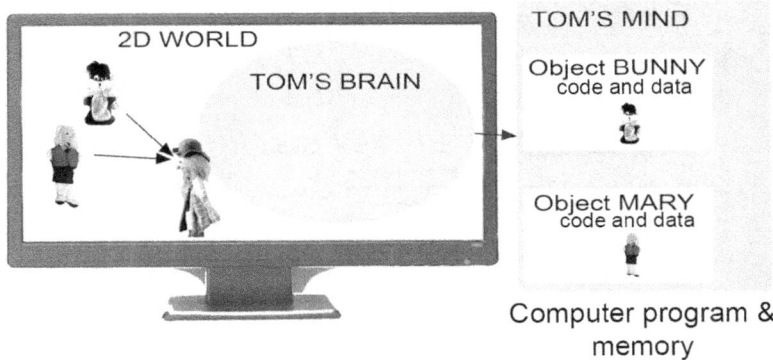

3D WORLD

Fig. 4.1 *Tom in our model observes Mary and Bunny.*

Same as oour brain in our three-dimensional world, *Tom*'s brain also has limited capacity and can register only a certain amount of data.

Evidently, storing and retrieving information in our model has to be organized and there has to exist some link between *Tom*'s 2D brain and the information stored on the computer.

In our 3D world exist programs called databases, designated to handle the storing and retrieving of information from computers. They use computer storage to accumulate the actual information passed to them and then they continually update a series of indexes, each pointing to a specific address where the particular information was deposited. These indexes are used to effortlessly retrieve stored information, referred to as object's *data*. In a database the series of indexes are called the *stack*, and since it occupies only a small amount of memory, it is much faster to search than the whole database structure.

The stack could be broadly compared to a wife's note on the table, instructing her husband returning from work where his dinner and slippers are.

When *Tom* sees *Bunny* and *Mary*, all the information about them is passed by *Tom*'s brain to a program running on the computer, i.e., *Tom's* mind. Program then stores this information in its database and creates an index pointing to the address where the information was deposited. This new index is then added to the series of already existing indexes in *Tom*'s mind, we would call it the *mind stack*, which is then copied to *Tom's* brain, into the *brain stack*.

Tom's mind stack is infinite, but the *brain stack* is limited. The new index is added on the top and if the *brain stack* is already full, the index at the bottom is discarded and all indexes move downwards.

The discarding of an index in the *brain stack* represents how we "forget things". If we ever encounter the same information again, the new index is created on the mind stack, then copied to the *brain stack* and we "remember" again.

Whenever *Tom* wishes to retrieve any information, i.e., "tries to remember that information", his brain finds first the index to that information from the *brain stack* and then retrieves it from the computer memory, i.e., from *Tom's mind stack*.

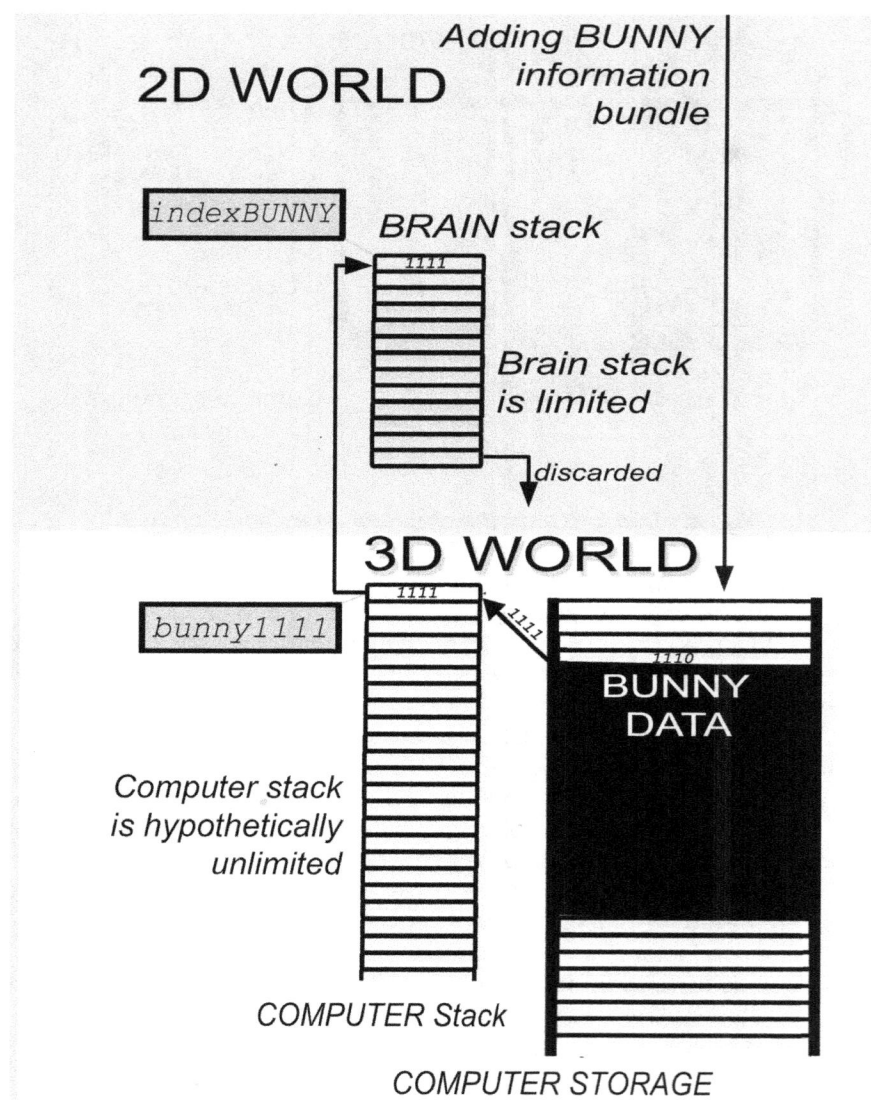

Fig. 4.2 *Tom in our model of our world observes Bunny.*

This scheme offers possible basic communication procedures between our brain and our mind. How the indexes are created and organized has to be programmed into the object's code stored in infinity, the same way as it is programmed in the computer program code, running on the PC.

Tom's "mind" in our model is stored on a computer in a three-dimensional world and not in his two-dimensional world. We could say that the computer memory where the mind of objects resides, contains:

Objects' code, which creates objects during program execution.

Objects' stored data, containing object's information.

Series of indexes to this data, i.e., the *mind stack*.

When an object is created, the object's *mind stack* is downloaded from the computer memory and the object's *brain stack* in the object's image is updated.

As long as an object is displayed, the object's *brain stack* exists.
As long as an object exists, the object's *mind stack* exists too.

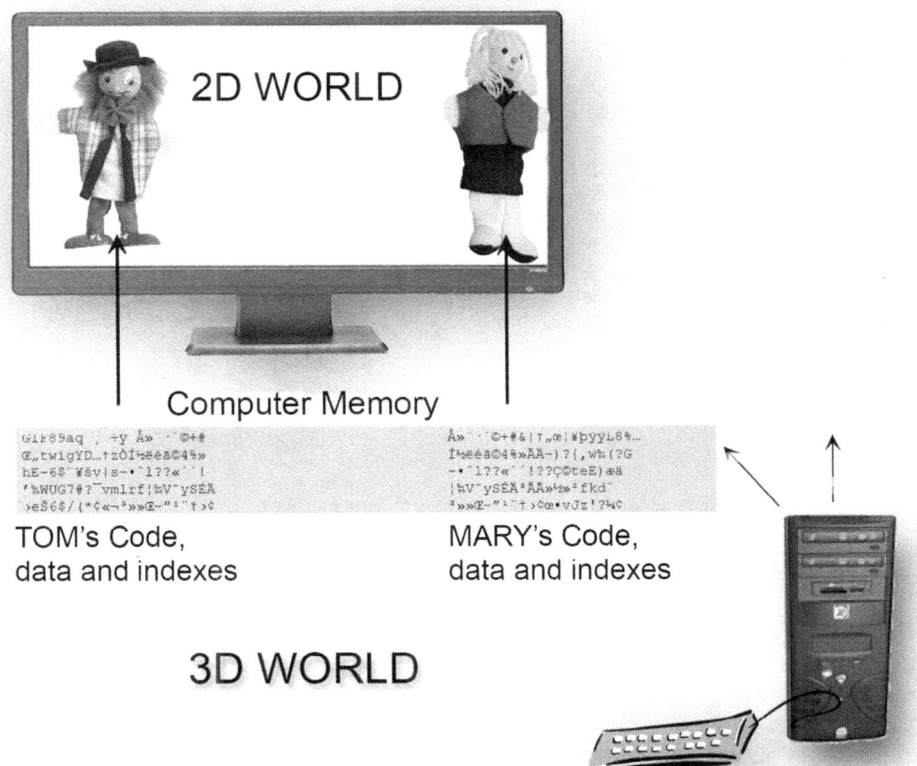

Fig. 4.3 *Tom and Mary existing in 2D world and their representation in 2D & 3D world .*

15

In the following example, we substitute in our model's two-dimensional world of computer screen by our three-dimensional world, and our three-dimensional world by infinity. As previously assumed, *Tom*'s two-dimensional brain contains only indexes, which implies that our brain in a three-dimensional world also contains only indexes. They are pointing to some addresses in our mind in infinity where our data is stored.

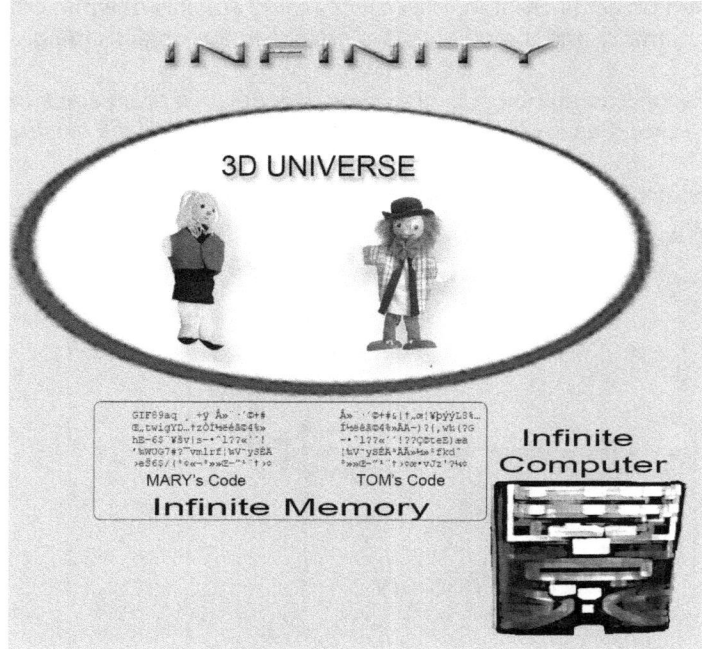

Fig. 4.4 *Representation of Tom's and Mary's, existing in 3D & infinity. worlds*

As we already stated, the capacity of the human brain is limited. It is believed that it has a maximum capacity of 2.5 million gigabytes of data. Although this is a large number, considering the amount of daily encounters with new objects and the length of human's life, evidently this capacity could not be adequate.

As with Tom's brain in a two-dimensional world, the human brain has to contain also the brain stack, with corresponding infinite mind stack and infinite data storage in our mind.

Obviously, the biggest part of the object's brain's capacity would be taken by the brain stack, which leaves a much smaller part of the brain to be used to perform all other tasks. This is the functional part of the brain and takes care of the running human body. It contains representation of abstract data like feelings, behavior constraints, instincts, consciousness, etc., and also it sends instructions to the body, like increasing the heart beat, experiencing feelings of pain, etc.

How our brain connects to an infinite mind is still a mystery. If we look at our model to explain this phenomenon, we find that Tom's brain, i.e., only a part of Tom's image, is connected to the computer by a series of wiring, conducting electric current.

Evidently, there is no wiring connecting our brain to the infinity, yet there has to be some connection. The only plausible explanation could be that the connection is maintained by the phenomenon we call "telepathy".[1]

It is still mysterious, but its existence is plausible since our brain definitely communicates with our mind.

During our sleep or when we are unconscious, this series of indexes disappears from our brain, i.e., our *brain stack* is empty, and when we wake up our brain starts downloading existing series of indexes directly from our *mind stack*. That could be supported by observations of the state of our unconsciousness during waking up: Initially we might not know what date it is, where we are, who we are, etc., but after only a short moment we usually remember it all.

Our brain also deteriorates, some indexes vanish and what we remembered before, we cannot remember any more. However, that information is still stored in our infinite mind, but it is not reachable from our brain. For example, when the brain's capacity deteriorates due to some illness, some indexes pointing to the stored information will be lost, but the stored information in our mind still remains. This has been proven already by many experiments with induced partial sleep and hypnosis. (We will discuss hypnosis in chapter 8. Hypnotism.)

The *brain stack* is perpetually being updated, i.e., some indexes are newly created, and because of the limited capacity of our brain, some indexes are also discarded. We then do not remember some particular information any more, but refreshing this information will create its new index on the *brain stack* again.

We could conclude that our finite brain is the connection to our infinite mind.

[1] *From Wikipedia : Telepathy (from the Greek τῆλε, tele meaning "distant" and πάθος/-πάθεια, pathos or -patheia meaning "feeling, perception, passion, affliction, experience")[3][4] is the purported vicarious transmission of information from one person's mind to another's without using any known human sensory channels or physical interaction. The term was first coined in 1882 by the classical scholar Frederic W. H. Myers,[5] a founder of the Society for Psychical Research (SPR),[6] and has remained more popular than the earlier expression thought-transference.[6][7]*

Telepathy experiments have historically been criticized for a lack of proper controls and repeatability. There is no good evidence that telepathy exists, and the topic is generally considered by the scientific community to be pseudoscience.
https://en.wikipedia.org/wiki/Telepathy

5. Images and Objects

From our model it is clear that when using a camera, connected to the computer as an *observer* of a three-dimensional cube, it is up to the computer program to form a two-dimensional image of this cube and display it on the computer display. That way we manage to display our three-dimensional object in the two-dimensional world of our model.

When three-dimensional *observers* are looking at the two-dimensional image of a cube, displayed on the computer screen, their brain will form in their mind either a two-dimensional or three-dimensional image. (Those *observers*, who already saw a three-dimensional cube before, will form in their mind a three-dimensional cube.)

When the same *observers* look at the three-dimensional cube, created in our world by computer in infinity, they will definitely form in their mind a three-dimensional image.

That is despite the fact that the images created by the human eyes are only two-dimensional. Fortunately, humans have two eyes, producing two independent images of the world around them. These images are then combined together by our brain to form a stereo vision, allowing us to sense the third dimension. The object created in our mind is then three-dimensional.

To support this statement, just look at some objects near you, then close your eyes and imagine one of the objects. Its image in your mind would be three-dimensional, despite the fact that it is only a combination of two, two-dimensional images, created by your own two eyes.

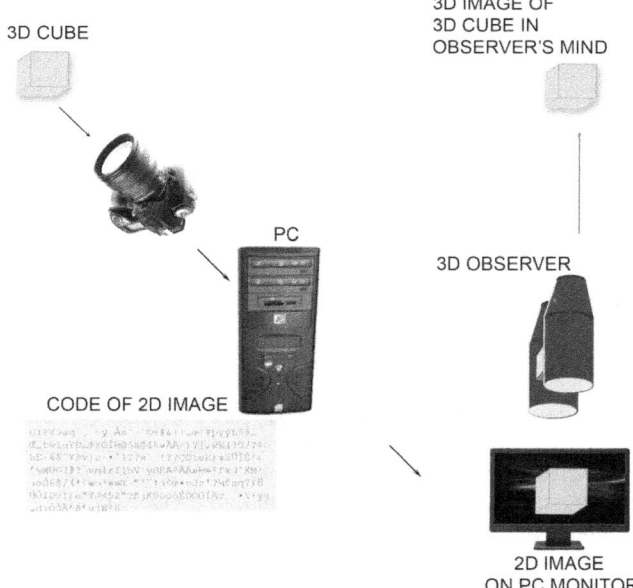

Fig. 5.1 *Transformation of three-dimensional physical object into computer memory, then to the display and finally to the observer's mind.*

18

As in our model, all objects in our world have to be created first, before they could be observed. That implies that a three-dimensional cube in our world has to be created before it could be seen. This is logical enough and it applies to all objects in our world, animals, plants, rocks, humans, etc.

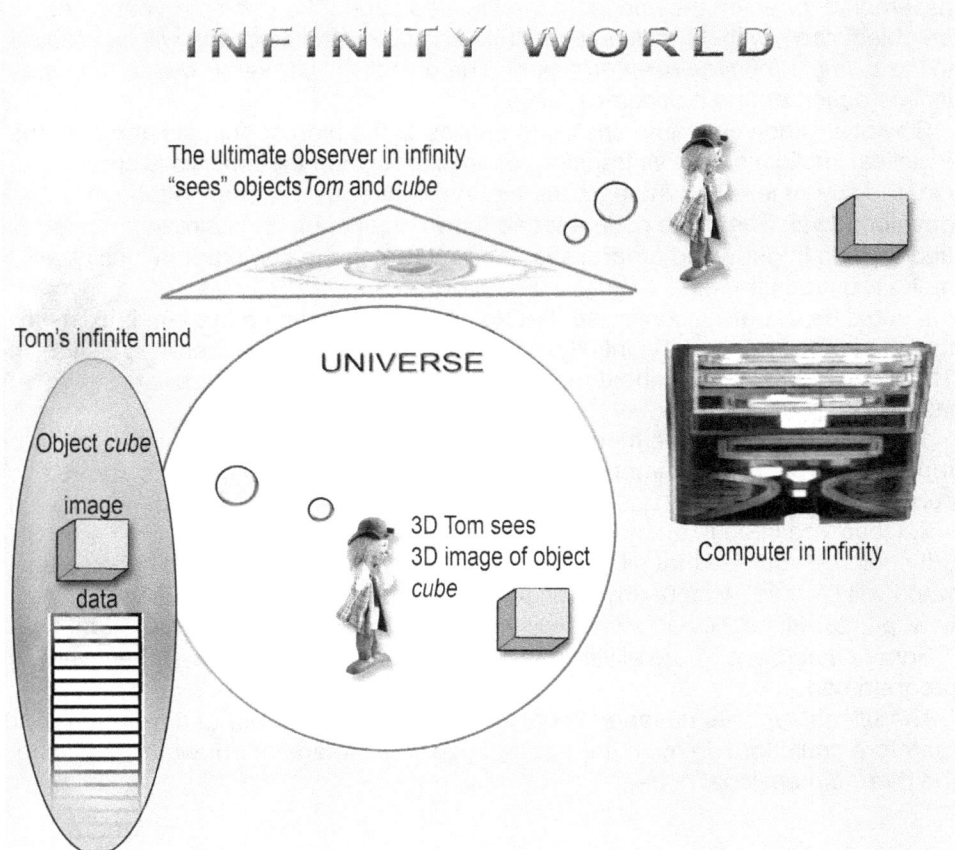

INFINITY WORLD

The ultimate observer in infinity "sees" objects *Tom* and *cube*

Tom's infinite mind

UNIVERSE

Object *cube*

image

data

3D Tom sees 3D image of object *cube*

Computer in infinity

Fig. 5.2 *The creation and transformation of physical objects in our world into infinity.*

In the example above, the computer in infinity creates the *universe* and objects *Tom* and *cube*. The *observer* in infinity "sees" both objects after they are created.

Because we have already proven that there is no light in infinity, the "seeing" is different in infinity from seeing in our world. It could be compared to "seeing" during our dreams. There we do not use our eyes and yet, we see objects in our dreams; some of them we think we have never seen before.

During our sleep our brain is partly switched off, i.e., the *brain stack* is empty, but the functional part of the brain still supports most of the bodily functions. While acquiring information stored in our mind in such a situation, we are bypassing our brain and communicating directly with our mind.

The same applies to the *ultimate infinite observer* - it connects directly to the mind of objects created in infinity.

It is evident that not all objects in our world are fully created by the infinite computer. Some of them were created as a result of interaction between already existing objects. For example, when Tom assembles a cube, he initially sees objects representing different parts of the cube, and only when they are assembled together, they become the finished cube. The cube then becomes an object itself, with its code in an infinite computer and its image will be created in the *ultimate infinite observer's* mind. The object will be "seen" by the *ultimate infinite observer* and it becomes "alive".

Obviously, the very same analogue applies to the birth of animals and humans. Practical implication of this transformation is that when the embryo is created, i.e., the sperm fertilises the egg, the embryo is already a human, although not developed yet. The basic code needed for an object human is already created and as time in our world progresses, the program in the computer in infinity will make required changes.

It could be therefore confirmed that an embryo is already a human. It exists in the mind of the *observer* in infinity and in our universe, and is therefore "alive". The same could be said about the object cube in our example. It exists in two instances:

1. Observed cube in the mind of the *ultimate infinite observer*, which is represented by a three-dimensional object cube in our universe; the observed cube is "alive".

2. Observed cube in the mind of a three-dimensional *observer*.

It could be decreed that all the images in our universe are alive, otherwise our world will be rigid, without any changes - humans will not move, rivers will not flow, plants will not flower, stones will not split, etc. Their lives differ, though.

Some objects are "more alive" than others, depending on the way they were programmed.

The *ultimate infinite observer* is not restricted by the number of dimensions and therefore could form in mind the three-dimensional image of our world, including the three-dimensional cube.

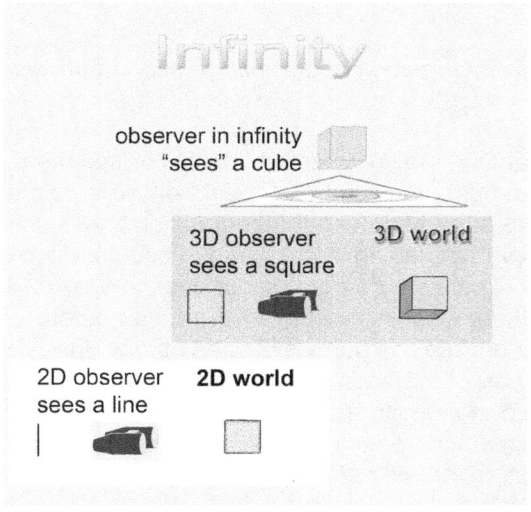

***Fig.** 5.3 The transition of images into objects*

Since three-dimensional *observers* in our universe can see two-dimensional images only, they create in their mind a three-dimensional cube, with the help of two superimposed, two-dimensional images.

The two-dimensional *observers*, displayed on the computer screen, observe only a one-dimensional image of the three-dimensional cube. They could see only a line, i.e., only one-dimensional image, no matter how many eyes they have.

Concluding, the progression from two-dimensional to three-dimensional world and then to infinity clearly demonstrates that by moving from the world of lower dimension to the world of higher dimensions and infinity, observed images become objects.

To refresh the concept of our model, we recap the basics:

Every object in our programming model consists of its code, stored in computer memory, and it could be also represented by its image, displayed on the computer monitor. Lines of object code are reachable only from our three-dimensional world and objects in our two-dimensional model have no meaningful means of changing the code.

There is a widely spread phenomenon called *self-programming,* claiming that an object during execution of the program could change some of its associated code. If that would be possible, then objects created in the computer during program execution would have to know how to program. However, this is not the case and programming could be done only by a programmer, who can write such a code. All that the created objects can do is to store and retrieve data and make some decisions, programmed to their code prior to the execution of the program. To imitate self-programming, all objects would have to obediently use some selected code, already prepared prior to running the program.

Judging by this model, two distinctive parts of every object, i.e., written code and object's displayed image, could be broadly compared to our soul and our body. Our body is then nothing but an image in the mind of the *ultimate infinite observer*, existing in infinity, and our soul is an object created from the lines of code and data, written and stored in a computer in the infinity world. Even a stone then has its soul, but it seems to be programmed differently from our soul..

Since in our model all objects have their associated code stored in the computer, it is obvious that they all have their soul, but not necessarily have their image displayed on a computer monitor. We do not see them, but we still know about their existence. We could have, for example, an object called a factory, which creates some other objects and their images.

Applied to our world, we could feel from many indications that there exist some objects, with their data and functions, that we cannot see. Like for example the Earth's gravity.

The existence of light makes it possible for *observers* in a three-dimensional world to see objects in a two-dimensional world. The transition between all objects in our world and the infinite mind does not need the light any more. There is the only one *ultimate infinite observer*, capable of "seeing" the very same way as we "see" in our dreams.

Time, as we know it, does not exist in infinity. Our mind is ruled by our *Subjective Time,* which during our life is constantly changing its rate of flow.

Einstein has already mentioned such relative time in his famous quote:

"When you are courting a nice girl, an hour seems like a second. When you sit on a red-hot cinder, a second seems like an hour. That's relativity."

The time our body is experiencing is different from the time experienced by our mind. Our body is not ruled by our *Subjective Time*, but by the time of our universe, i.e., the *Universal Time*[1], i.e., *Newtonian time*[2], which has a constant rate of flow and rules our presence in our three-dimensional world. Basically, our world is only a broadcast of events, happening in the infinite world of our mind. It is similar to a school blackboard, which depicts the results of processes taking place in teachers' and pupils' minds.

There is a very notable difference, though. Unlike objects on the blackboard, we are objects equipped with self-awareness and we feel the existence of our body in this three-dimensional world.

6. Life

In our two-dimensional model, objects "live" as long as they are created, activated and possibly displayed on the screen. That requires that every object must be defined in the program code of our model. For example, object *Tom* is defined by a code segment "*Tom*", which is a subclass of class "HUMAN".

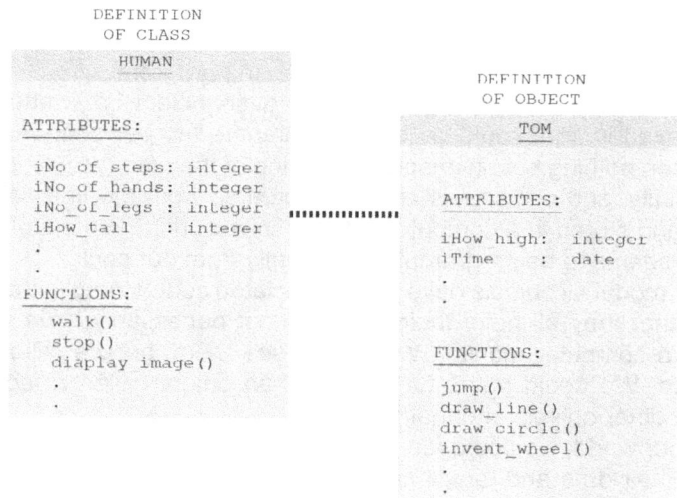

Fig. 6.1 *To create object Tom, the program uses class HUMAN as a blueprint.*

[1]This statement is contrary to the *special theory of relativity*, developed by *Albert Einstein*, approx. hundred years ago. The detailed analysis with supporting logical and mathematical arguments is in the Attachments section.

[2] Defined many centuries ago by *Isaac Newton*.

In our model the program creates an object in computer memory and if this object also has a code segment defining its image, it creates the object's image on the monitor or projection screen. Not all objects have a displayed image and it depends on how a particular object was programmed.

During the execution of the program, an existing object reacts to occurring events and messages addressed to it, and then changes its attributes accordingly. For example, the attribute *iTime* could be reserved for ever changing *SubjectiveTime* of the object.

These attributes are stored in computer memory and retrieved whenever the object needs to use them. Such continuous updating of attributes could be compared in our world to "learning and gaining experience while we are alive".

During execution of the program, different objects react with each other and the results of such interactions could also reflect on computer display. However, the images on the display are only the results of processes running in computer memory, of which two-dimensional object *Tom*'s memory is an integral part.

Display reflects what is currently processed and sends data to the program. For example, it sends a signal when an object moves outside the visible screen.

Display by itself does not program any processes. Even a display with touch screens acts only as a sensor, sending data to the program, where it is processed by the computer and not by the screen.

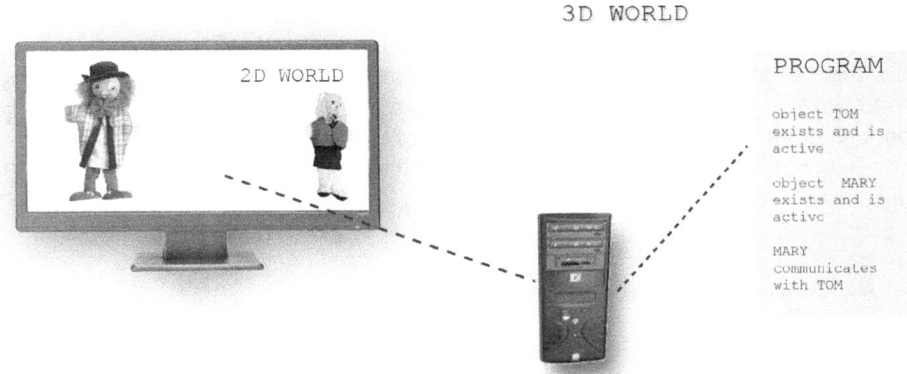

Fig. 6.2 Objects Tom and Mary exist in the computer's memory and on display are represented by their images. They could communicate and update the values of their respective attributes.

Our existence could be compared to the existence of objects in our model. They exist as long as the program does not destroy them. That could happen in three possible ways, and all of them could be valid:

1. Program will destroy *Tom's* displayed image and will remove his object's code and data from the program's memory. Object *Tom* will cease to exist.

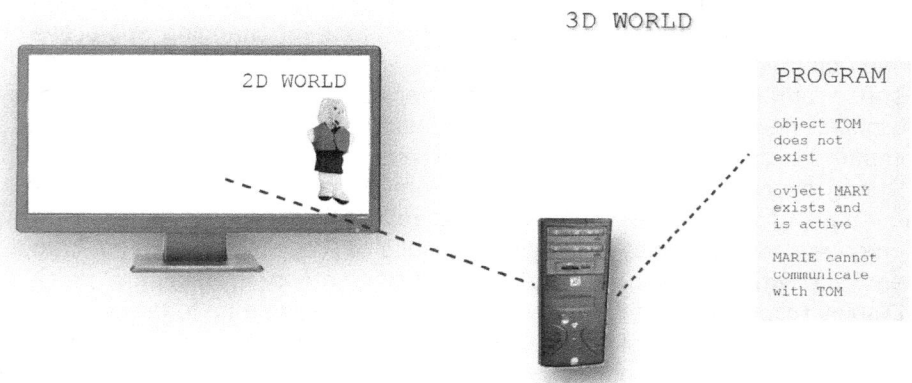

Fig. 6.3 *Object Tom and his image do not exist any more. Mary therefore, does not see Tom and cannot communicate with him.*

2. In the second alternative, the program will remove the object's code and data from the program's memory, but its image will not be removed. It will become a "dead" image, unable to move and react to messages sent to the object. Object *Tom* will cease to exist and the part of the program's memory where the object's code was stored, will be marked as free and it will be available to be used for other purposes.

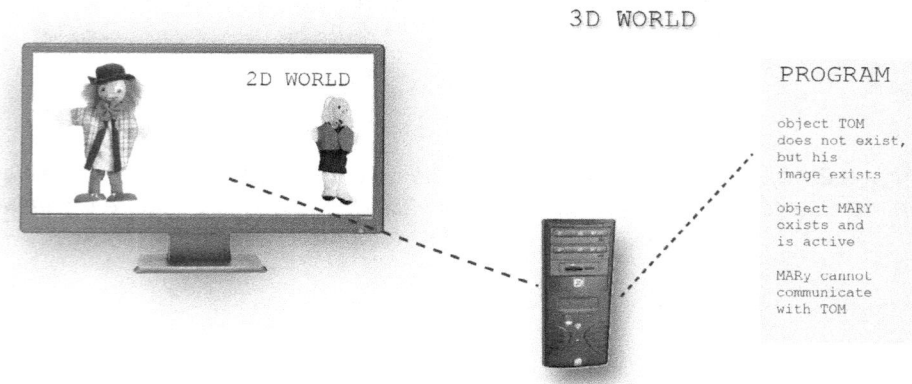

Fig. 6.4 *Object Tom does not exist in the program's memory, but his image on the computer's display still temporarily exists. Mary will "see" Tom, but will be unable to communicate with him.*

3. In the third alternative, the program will destroy *Tom's* image and code, but *Tom's* data will stay in the program's memory. It will not be accessible by program's functions and only the *ultimate infinite observer,* and possibly objects existing only in infinity could access it.

If required, the object code and image could be created again with an empty data segment. It does not have to be exactly the same image as the previous one. (Object could be programmed to have more than one blueprint for its image.)

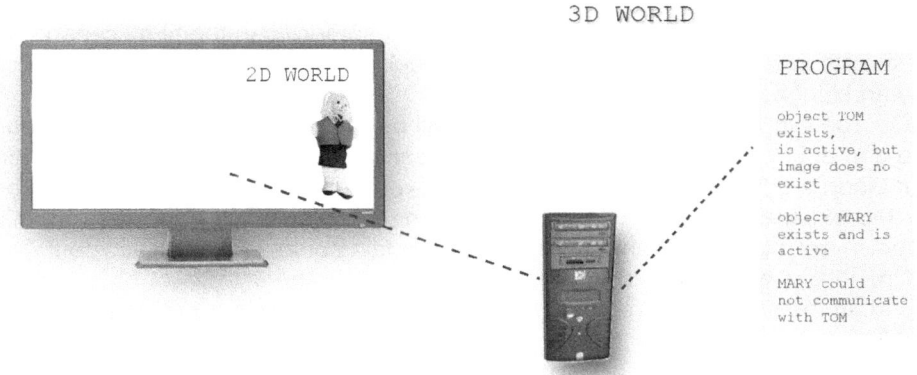

Fig. *6.5* Object Tom's data exists in the program's memory, but its displayed image and code was destroyed. For Mary, Tom is "dead".

In all these alternatives, the program's code for object *HUMAN* does not change. This code was used to create the object *Tom* and it still exists. If required, the program could create a new object *Tom.*

Described scenarios could be compared with similar scenarios, existing in our world:

In the first scenario, a person dies, i.e., body, mind, running code and all the collected data in the memory of the infinite computer running our world will be destroyed.

In the second scenario, the person's body will still exist, but the person's mind, code and data collected during the person's life will be destroyed and the person's soul dies.

Both scenarios do not make any sense. If any one of them is true, then our existence in this world is meaningless. Everything that we learnt will be lost!

You can choose any one of these scenarios, depending on what you believe. For me, the only third scenario is the most probable. Although persons' bodies will cease to exist, persons' minds will still prevail. These persons will not be represented in our world by their bodies, but their minds will still exist in an *infinite world.*

As long as we believe that our life has some meaning, then this alternative is closest to our situation. It means that after death, a person's mind still exists in an *infinite world.*

Consider again that our programming model is a game where we could create an object, according to our wishes. For example, I could create an object George, with some of my specific attributes. Then I guide George through all different situations, which the program's rules and scenario will allow. The goal would be for George to live right up to the set life span and learn from his encounters as much as possible.

While playing, my mind in infinity and George's mind in the computer fuse into one and I could easily forget about my existence anywhere else, but in this two-dimensional game. That happens, for example, when we are intensively watching some movie, which induces in our mind feelings specific to the film scenario. We associate with characters in the movie, we feel joy, anger, mental pain and we can cry and laugh.

George's input into our common mind will be based on his encounters through the game. For example, George drives his car too fast and the car hits a barrier.

If George gets killed, the game for me is over and I return to my world, enriched by the experience I had while playing the game. George is gone, but his mind stays in my mind. I still remember that driving fast could lead to an accident.

Previously, we set the objects in our model to be the images of objects in our world. In the same way we could exist as images of objects, existing in an *infinite world*.

Obviously, only the already mentioned third alternative offers an explanation, which gives a logical meaning to our life. Then, of course, there are many other forms of life, and should our model also apply to them, then we would have to completely change our beliefs and behaviour.

"My religion consists of a humble admiration of the illimitable superior spirit who reveals himself in the slight details we are able to perceive with our frail and feeble mind."
Albert Einstein

7. Most Likely Scenario

So far, we concluded that our universe is not infinite, but our mind is. We also assumed that our brain is connecting our existence in this three-dimensional world with our mind in an infinite world. We derived this assumption logically, since there is nothing else that could have this functionality.

We have this situation already depicted in our model where the object's brain exists on the computer screen in a two-dimensional world, and the object's mind exists in computer memory, in our three-dimensional world. We could enhance the program of our model and turn it into a sophisticated game where objects interact with each other and their surroundings, and respond to commands sent to them via keyboard by one or more players of this game.

Fig. 7.1 George in our world plays the game on a computer.
2D Images of Tom and Mary exist on the computer screen.
Their 3D images exist in George's mind, i.e., in infinity.

In the figure above, George is playing a computer game, in which displayed objects on the screen have only limited capabilities. For the game to resemble real people in our world, we would have to, at least hypothetically, add many more attributes and functions. For example, we would add the ability to make decisions, completely independent of the player's will. We could also add functions, reacting to different *Tom*'s and *Mary*'s sensors. They would be able to "see", "smell", "hear", etc. All that could be added with some advanced programming and sophisticated computer hardware.

With these enhanced capabilities, *Tom* and *Mary* could resemble real people living in our 3D world. Because these objects are created in George's mind, *Tom* and *Mary* objects exist in the *infinite world*.

During the game, George's mind could become concentrated only on one object, for example *Tom*, and that way it will become *Tom* himself, having all the benefits of existence in George's mind.

Imagine yourself playing some adventurous game where you select a figure representing you during the game. You become immersed in the intricatenesses of the game and soon forget about your surroundings. You might forget about buying a bottle of milk on your way home, as directed by your partner, or forget about your real existence entirely. During the game you become Tom, existing in a two-dimensional world of a computer screen.

We can even go one step further and change the program to create pseudo three-dimensional images of all objects and then replace the computer screen with three-dimensional goggles. You will not see your three-dimensional world around you any more, since you will see only the world created by the program.

Lastly, but not last, Tom's image could be replaced by your hologram image and a set of sensors and electrodes could be placed on your body. They will be sending signals representing some feelings of your body to the computer program, and vice versa.

If all these add-ons could be so advanced that they could even interact with your brain, then our model will be very close to our real world.

After all, we have already concluded that we are three-dimensional images in the mind of the *ultimate infinite observer* in infinity. Instead of hologram images of ourselves in our mind, in the mind the *ultimate infinite observer* our images are made of earth's elements. Since that world is infinite, there also exist infinite possibilities beyond our imagination.

I might even suggest that we exist in an infinite world and in our three-dimensional world we are only three-dimensional images of ourselves. We could be subjected to an induced sleep and exposed to some computer program, which will create this three-dimensional world. Our brain will contain an empty brain stack, freshly created when we were born, and thus it will limit our knowledge only to what we learnt in this three-dimensional world.

We do not have any knowledge about the infinite world and initially, about our three-dimensional world either. We are afraid of dying and we do not know what happens after we die. This is purposely included in the code, from which we were created, and its purpose is preventing us from ending our lives by our own, individual decision.

To end with a possible scenario for the existence of our world, I believe we could be even a penal colony, to which we were sent for our misbehaviour in the real, infinite world. After all, there is nothing existing in this world that we could take with us to the infinite world. What we could take is the knowledge and feelings we have acquired during our life here. If we became a better person, then the mission succeeded. If not, we might be reincarnated and serve the "life sentence" again.

After all, this belief that humans *start to live when they can live outside themself,* was already expressed by *Albert Einstein.*

Remember, our world is only a finite bubble, created in the infinite world!

8. Hypnotism

All that we are is the result of what we have thought. (Buddha)

By using our hypothetical model, we have already derived a simple, yet depicting representation of our world. This model could be used further to unravel some still existing puzzles, like for example hypnotism,.

Despite the long history of hypnotism, this ancient phenomenon already mentioned in the Bible, is even now still shrouded in mystery. Christ was able to cure sick people and the ancient civilizations used hypnotism to restore health and suppress pain.

There exist many tangible records of such practice and one of the most remarkable devotees of hypnotism was doctor Franz Anton Mesmer[1], who lived in Vienna during eighteen century. The English language even invented words *'mesmerize'* and *'mesmerism'*, derived from his name.

His most known achievement happened in 1777, when Maria Theresa Paradise, infant prodigy and blind pianist, recovered her sight after his treatment. At that time she was already under the care of Dr. Von Stoerk, best oculist in Europe and apparently, for ten years there was not improvement.

The very recent and revealing demonstration of hypnosis I have encountered was in one episode of the excellent documentary series Mythbusters, created in 2007.

The team decided to investigate whether the hypnosis should be considered a myth or not. Three of the team, Grant, Kari and Tory were firstly tested for suitability and then they were unknowingly exposed to an arranged live episode, involving actors: Two actors delivered a package, had a brief and even heated conversation and then they left.

Each member of the team then received a questionnaire, relating to that played episode, and they were asked to answer as many questions as they could remember.

During the second part of the experiment, the team members were hypnotized, and only then received a new questionnaire. While hypnotized, they answered the same questions again.

The answers in the first questionnaire were only a few noticeable observations, but in the second questionnaire, done under hypnosis, the answers were more detailed. For example, they remembered the displayed name on men's name tags and the description of the tattoo, one of the men had on his neck. The conclusion of their experiment was that hypnosis is for real and it is not a myth.

When the team members were hypnotized, they seem to fall into an induced, different state of mind, often referred to as subconsciousness. They were still able to answer any questions, but their spiritual presence in this world was somehow altered.

To find an explanation to this experiment, we could try our model and let *Tom* hypnotize *Mary*. Her brain becomes partly blocked, and fully responsive is only her mind, residing in our three-dimensional world.

[1] *'HYPNOTISM AND THE POWER WITHIN'* by Dr. S. J. Van Pelt

Outside the hypnotised state, when *Tom* is talking to *Mary*, he is communicating through her brain, which in turn communicates with her mind. During hypnosis, the *brain stack* in *Mary's* brain is blocked and *Mary's* brain accesses stored information in *Mary's* mind directly through her *mind stack*.

We have already derived that the *mind stack* is limitless. The *brain stack,* on the other hand, is finite and its size is limited by the size of the brain.

Inevitably, not all the addresses of stored information, ever registered by the *mind stack,* are necessarily registered by the *brain stack*.

That explains why under hypnosis, the team members were able to remember more details from the staged episode.

3D WORLD

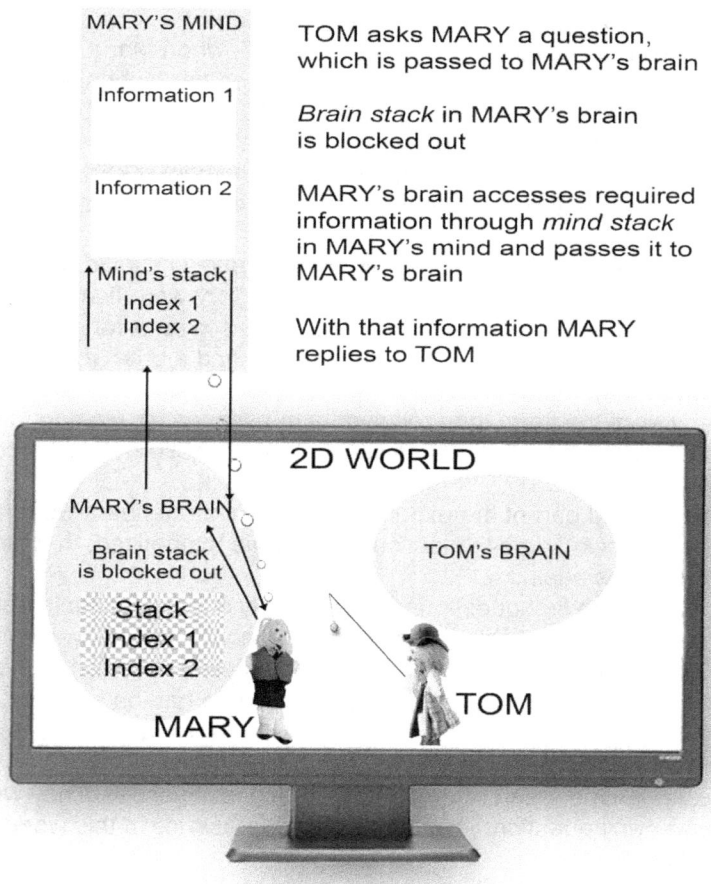

MARY'S MIND

Information 1

Information 2

↑ Mind's stack
   Index 1
   Index 2

TOM asks MARY a question, which is passed to MARY's brain

Brain stack in MARY's brain is blocked out

MARY's brain accesses required information through *mind stack* in MARY's mind and passes it to MARY's brain

With that information MARY replies to TOM

2D WORLD

MARY's BRAIN

Brain stack is blocked out

Stack
Index 1
Index 2

TOM's BRAIN

MARY

TOM

Fig. 8.1 *Tom communicates with hypnotized Mary.*

The results obtained by Mythbusters experiment suggest that the hypnotized team members' brain communicates directly with their mind stack, same as does Mary's brain in our model.

Similar to hypnosis is anaesthesia, used to induce the state of low or no responsiveness at all. The difference between anaesthesia and hypnotic stage is the intensity of reduced activity of the brain. The anaesthesia induces deep sleep, during which the person does not respond to outside events at all. On the other hand, hypnosis is more likely compatible with a less intensive stage of awakening from anaesthesia.

At this stage of awakening, there is a mischievous practice to ask the affected patient some very intimate questions. Usually, the patient will reveal all what is otherwise a closely guarded secret. In a fully awakened state, the personal feelings and moral constraints, being bodily attributes stored in the human brain and not in our mind, will prevent this from happening.

Good examples are the feelings of joy, guilt, remorse, pain, etc. These feelings and constraints are part of our body, i.e., the body feels pain, noise, movements, etc. They are specific to a particular, ever-changing point in the *Universal Time*, and since this time does not exist in infinity where our mind exists, bodily feelings and constraints cannot be a part of our mind.

They are relevant only to the brain and when the mind is being accessed directly through the mind stack, these constraints and feelings are not registered there.

There is definitely a separation of bodily functions and the person's mind. We can conclude that hypnosis partly bypasses a person's brain and connects the hypnotized persons directly to their mind in an infinite world.

Hypnosis is therefore the only small and temporary window into the infinite world of our mind.

9. Our Dreams

The existence of our mind in an *infinite world* creates an inevitable question: How does our mind exist in such a world? The answer to that could be found by answering another question: How do we exist in our dreams?

We experience dreams during our sleep and they could be represented by a mysterious combination of imagination, knowledge and events. We do not use our eyes to see in our dreams or our ears and vocal cords to communicate.

Our dreams are like an interactive movie where we are the main actor. We "see" our dreams through our partially subdued brain and sometimes we even intensively feel all associated emotions. The time involved in our dreams is our *SubjectiveTime*, with an uncertain rate of flow. As during hypnosis, our dreams do not access our brain for stored information, but connect directly to our mind.

While dreaming during a deep sleep, we usually do not feel our body and actually we do not feel anything except some emotions, which are the products of our mind. Sometimes during the dream, some of the created emotions are automatically passed from our mind to our subdued brain, which in turn obediently creates some bodily function.

For example, completely without our consent we cry or laugh in our dreams, and sometimes our body does the same. When we wake up from our dream, our direct connection to the mind becomes broken and is redirected to the *brain stack* again. For the duration of this change we still remember the entire dream, but soon after we forget what the entire dream was about. In our dreams we see without using eyes, listen without using ears and communicate without using vocal cords.

Many could argue that dreams simply reflect unintentional processes, mistakenly being carried out by our brain during sleep. Contrary to that, I believe there is an existing reality involved in our dreams. For example, many people are having dreams, during which they speak one or more languages they have learnt in their life so far. That knowledge is real and not a pure fantasy created by the brain, and as we already deduced, this knowledge is stored in our mind.

Sleepwalking would be another example of our body communicating directly with our mind. After waking up, persons do not remember at all they were walking, and do not remember anything they have encountered during that time. Their brain was subdued and the *brain stack* did not register any events.

During sleep or under hypnosis, the *brain stack* is disabled and the brain communicates directly with our mind. As a consequence of that, we could access the information stored in our mind, but unfortunately only accidently, and we almost immediately forget it.

There is no plausible explanation of what the dreams are and there are many myths surrounding them. Some people believe that the future could be predicted by analyzing dreams. If they are correct, then the future is already established and is defined somewhere. But is it possible?

To investigate it, we again compare this situation in our world to our digital model. For example, some time in the future *Tom* and *Mary* meet in a restaurant. They both would have to be programmed to do so, well before that happens. When we add more objects, actions of every one of them would have to be synchronized with actions of all other objects.

Such a scenario will be not just impossible to program, but all the decisions of all objects, right from the beginning, would have to be also predefined. That would put our model on par with some passive game or movie. The decision-making will be taken away from individual objects and our model will not represent our world any more. Obviously, the future in our model does not exist.

That does not mean that there is no future in the infinite world! The existence of a pre-defined future could be investigated by considering the existence of the *present time* in our world. When looking at the watch, we cannot say that the exact time now is our present time. As soon as we say it, the hand on the watch already points to a different time and the present time has already passed. No matter how fast we measure our time, any time existing in our world is always the past. The present time exists only in infinity and is therefore infinite. To look to the future, we would have to cross that infinity partition, which in our finite world is impossible.

In our model and in our world, the future is obviously beyond our reach. That does not eliminate for us the existence of the future in the infinite world, though. In infinity exists only our *Subjective Time*, and therefore our future there exists and is represented by *immortality*.

10. Parallel Universes

This concept pops up from time to time in many media, and somehow I always missed the plausible explanation of what the parallel universes actually are.

Using our model, let's imagine that there are more *observers* looking at the computer screen where *Tom* is displayed. When displayed *Tom*'s object is being observed by more than one *observer*, the same *Tom* object exists in the mind of each *observer*. All objects *Tom* form parallel objects in parallel universes, and all objects and universes are the same.

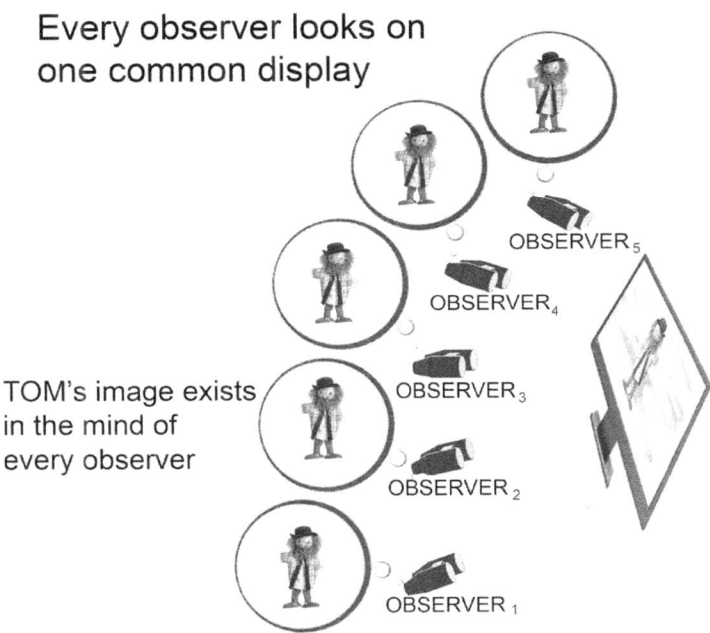

Every observer looks on one common display

TOM's image exists in the mind of every observer

OBSERVER 5
OBSERVER 4
OBSERVER 3
OBSERVER 2
OBSERVER 1

Fig. 10.1 Displayed object Tom is being observed by more than one observer.

From the picture above it is obvious that *Tom*s' image in *observers'* mind exists only if *Tom*'s image on the screen is observed. What is also obvious is that *Tom*'s image exists in more instances.

What also applies is that if anything on the screen changes, these changes propagate instantly to all instances of that object, created in each *observer's* mind.

This scenario nicely fits into the physics of *quantum mechanics*, which deals with miniature dimensions in the subatomic world of atoms, electrons and similarly small particles. It is a complex science and American physicist *Richard Feynman* once characterized it:

'*I think I can safely say that nobody understands quantum mechanics.*'

The picture 10.1 clearly demonstrates that our model already explains three basic postulates of *quantum mechanics*:

1. Object exists only when the object is observed.

2. If one object is observed by more than one *observer*, then this object exists in more identical instances. This is in agreement with the principle of *quantum mechanics* called *quantum superposition*.

3. Should anything change on the observed object, then this change is instantly propagated to all instances of that object. This is in agreement with the principle of *quantum mechanics* called *quantum entanglement*.

Since the concepts of *quantum mechanics* and our abstract digital model are independent entities, these shared attributes justify our belief in the validity of our model.

We do not know if parallel universes exist, and if they do, we do not know how many there are. It is obvious from our model that for each possible instance of the universe, there can be only one *observer*, i.e., one god.

While we are alive, the object consisting of our image and our mind exists in the infinite mind of this *observer*.

"I want to know God's thoughts… the rest are details."
Albert Einstein

11. Nature and Hidden Constants

The Gravitational constant $\underline{G}$ = 6.67x10^{-11} N m^2/kg^2 is for example just one of the hidden constants, existing in our world. Besides these constants, there exist also laws governing our world, like for example, four laws of thermodynamics. The question is, how such definitions and laws were defined?

Nobody from our world has invented such laws and nobody defined such constants. Their existence was found experimentally and their values were observed and calculated. Definitely, there must be even more laws of nature and more defined constants, which we have not discovered. Yet, they are ruling our world and we could sense their existence.

One such a mysterious phenomenon is randomness of events, which could be demonstrated in the following example:

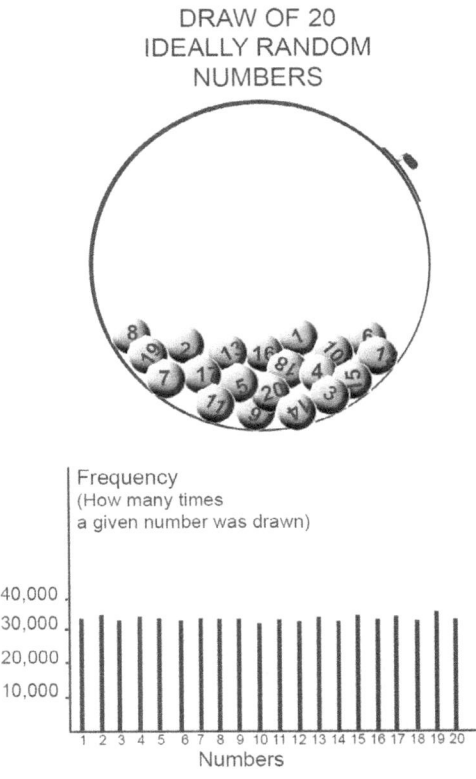

Fig. 11.1 *Balls, marked by numbers 1 to 20, are randomly drawn by ideally random apparatus. The draws of such apparatus are not influenced by any physical imperfections.*

During the experiment, after each draw we add a small bead to the tube representing the drawn number. For example, when we draw number 5, one bead is dropped into a tube marked "5". Then the ball is returned into the drum and drawing continues.

If we repeat this process, say 30 000 times, we will discover that the number of beads in the tubes, representing frequencies of draw of individual numbers, is increasing at almost an even rate. It does not happen that one tube will be filled considerably more than some other.

Even if this is not proof that there is a law governing the randomness of these draws, it is an indication that some laws must be governing the functions of this apparatus.

Here is another example of one peculiar situation:

Fig. 11.2 If the number "3" appears on the top after three successful castings - would you bet on that number again?

Should we look at this scenario logically, with every new cast, the number "3" has the same chance to be cast as all the other numbers. However, there is something urging us to believe that its chances are getting smaller.

From these two experiments we should not be burned at stake for believing that there is a law governing randomness in our world. To ascertain that, we can use our programming model and try to program similar mechanisms into its code.

Firstly, we should define randomness. When the numbers are drawn or stay on top when dice is cast, their series must not form a repeating sequence and the frequency of their draws must keep parity with all the other numbers.

To achieve this, with every draw we would check all the numbers drawn so far, and make sure that the new number does not form a repeating sequence.

To do that would require keeping of a database where we store all numbers already drawn. After every new draw, we would check this database for a repeating sequence.

If we find that there is no sequence being formed, then we proceed further and compare all the frequencies of drawn numbers. We could set for example, the allowable percentage of possible difference to 2%. If the frequency of the newly drawn number exceeds that, it will be rejected and a new draw would start. When both criteria are satisfied, the drawn number will be accepted.

At the start, this process will be fairly simple, but with the increasing number of draws it will become more and more complex. Drawn numbers will form combinations without repetition = $n!/ (k! \times (n-k)!)$

Example:
Two dices form two-numbered combinations (k) from 6 numbers (n) without repetition

$$6! /(2! \times (6-2)!) = 720/(2 \times 24) = 720/48 = 15$$

Let's consider another process, for example some typical lotto draw. After the first draw, the number of all different combinations to be checked is 3,838,380. (Six-numbered group (k = 6) out of 40 (n = 40) without repetition:

$$40! / (6! \times (40\text{-}6)!) = 3{,}838{,}380$$

With an increasing number of draws, the program would have to check an astronomical number of combinations and our random function will soon collapse.

Our model in our three-dimensional world is therefore not capable of ensuring the randomness of any function. Yet, we can see the results of such processes in our world, like for example the process of forming a human face. The explanation is that the code of such a process is not a part of our world, but it is part of a code, running our world from infinity. The creator(s) of our world used code that we cannot duplicate.

What we can duplicate in our program are hidden constants and some basic laws. In our model we call them "global constants and functions" and as such, they are valid for the whole program and all created objects. Their values cannot be changed and they are established at the start of the program and removed only when the program ends.

This comparison with our model explains the existence of hidden constants in our world - they were simply defined and are unchangeable. We have discovered many such constants, but there is still possibility that there are many more, waiting to be discovered.

We could now state that the laws of nature and hidden constants were defined when our code was created, even before our world existed. These laws and values of defined constants are set and we could not change them from our world. Some, like for example the constant of speed of light in different substances, are built into our code, which was created before we were created. They have not evolved.

"Everything should be made as simple as possible, but not simpler."
Albert Einstein

12. Living Model

Based on the assumption that we are just a sophisticated simulation of an infinite world, we are inevitably subjected to the existing environment of such a simulation. This simulation could be even a game, with its rules and goals. Not surprisingly, this game, which we call our "life" is not void of excitement and the scenario for this game includes a never-ending struggle between the good and evil, between the truth and lie, between the beauty and ugliness, etc., etc. There cannot be peace in our world, since this is not part of this game's goals.

If we try to look for a possible explanation of our life by comparing it with our model, we would not find any definite answers. Our model does not have the knowledge, which created our world.

We should then look around us and find some already existing models. There is a great probability that if some event in our world had already happened, it might happen again.

We could find some simple model, or at least a model simple to understand, and apply it to our civilization. The imposed condition on our new model is that it should have the same basic parameters as our civilization has. Namely, it has to be some population of living objects, with the ability to live and multiply. Other desirable parameters would be a suitable living space and availability of resources to be consumed.

When limiting ourselves to such simple requirements, all we have to do is to look at the living world around us. We would give our preferences to a population of objects with a short life span, since we do not wish to wait long to see any results.

One ideal example of such civilization of living objects, being everywhere, are microorganisms. The most suitable for our model would be the population of bacteria producing alcohol. We can create such a population by simply filling a jar with some fruit and water, and then add some bacteria culture, available to buy from a shop catering for home-brew enthusiasts. This new population has to be kept at room temperature and bacteria will start to consume the sugar in fruit. As with every living object, bacteria will also automatically produce some by-products, namely carbon dioxide and ethyl alcohol. The first is a gas and it will escape to the atmosphere, but the alcohol will stay dissolved in the water where bacteria live. Initially, bacteria will have no shortage of resources or space, and therefore will uncontrollably multiply.

The problem bacteria have is that the alcohol they produce is poisoning them. When they produce enough alcohol and the concentration increases to a level they cannot tolerate any more (approx. 17%), they start dying.

They were not programmed to realize the gravity of their situation and they will therefore proceed towards their own destruction regardless. Soon they all die and their place in this world will be taken by some other bacteria culture. Usually, bacteria producing vinegar will take over and we will have a new model to observe.

So far, all that corresponds to what humans are doing on this planet. Only our model proceeds to its end at a much greater pace.

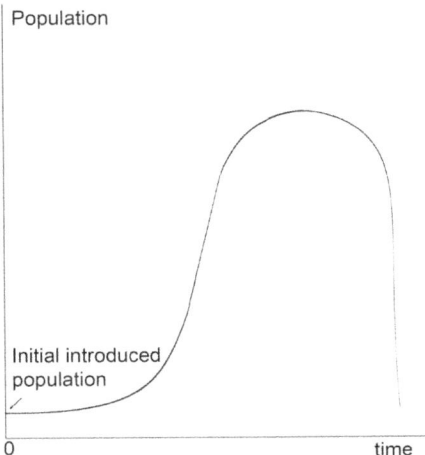

Fig. 12.1 *This graph is an approximation of a life-cycle of some typical bacteria culture. The position on the top of the curve represents either depletion of resources or unacceptably high concentration of poisonous by-products. In that stage, bacteria start dying faster than they are reproducing.*

The bacteria culture's problem is also one of the problems, so typical for our civilization. At the beginning, humans found themselves in a favorable environment with plenty of resources. They started to consume these resources and started to reproduce themselves. During this process they also produced and released into their environment poisonous by-products. Unfortunately, same as bacteria, humans have only limited resources and limited living space.

In the beginning of our civilization, the population density was low and epidemics and permanent wars kept the human population at sustainable levels. Also the concentration of poisonous by-products was low and all living objects coexisted in natural balance.

Then advances in science eliminated major epidemics and new technologies enabled humans to produce more and more food, to support a fast growing population. That brought with it a sharp increase in demand for resources and living space, and a sharp increase in concentration of poisonous by-products.

So far, there is no consensus amongst nations on how to keep population growth on a sustainable level and how to force industries to stop producing poisonous by-products.

This situation corresponds very fittingly to our bacteria model and should our civilization not break out of this vicious circle, we would follow the same fate of bacteria in our model. Since this scenario for our bacteria model was written in the code, used in creating and running our world, we have no reasons to believe that humans could be an exception and follow a different path.

"Technological progress is like an axe in the hands of a pathological criminal."
Albert Einstein

Book Summary

Considering our world as being only an image in the mind of an *observer* in infinity, explains nicely the paradox of something existing only if it is observed, and it explains why something could exist simultaneously in more than one place (*quantum superposition*). It also explains why any changes in the object's properties reflect immediately on all instances of that object (*quantum entanglement*) and it allows for the existence of parallel universes.

Objects in a two-dimensional world, when observed by an *observer* in a three-dimensional world, become images in the infinite mind of that *observer*. The same applies to objects in our world - there are three-dimensional images formed in the infinite mind of an *observer*, existing in another, infinite world.

The time in our world is the *Universal Time*, with the constant rate of flow, pointing from the past to the infinite present and future. There is no present time or future in our world, since both lie in infinity, which in our world does not exist.

Our mind is ruled by our *SubjectiveTime,* existing in infinity, which could have a different rate of flow for different objects. This rate could be affected by many factors, like for example the frequency of interesting events, sleep, etc. The *SubjectiveTime* exists only in our mind and has no bearing on events, progressing outside our mind at a pace set by the *Universal Time*.

Laws of nature and hidden constants support the belief that our world is a computer simulation, designed and programmed well before our world was created. These laws and values of defined constants are set, we could not change them and they have not evolved. We have already discovered some, like for example, four laws of thermodynamics, but the existence of some other we could only guess. Randomness of processes in our world is one example of such possible, undiscovered law.

Our programming model has simplified our perception of the universe. The logic of our model makes it easier for us to understand the beginning of our universe and also its possible ending. Unfortunately, it will still not answer many other questions, like for example, about the existence ot the infinite world?

It is obvious that in our three-dimensional world we will be looking in vain for such an answer. Exactly as *observers* in our two-dimensional model would be in vain looking for the answer to the same question in their world. It is we, *creators* of our model, who are the only to know the answer to their question. I believe that only the *creator(s)* of our three-dimensional world know the answer to our question.

There is a fundamental difference between the *creator* and *observer* of our world. There could be many *creators*, yet there is only one *observer* of our particular word. Should we call the *observer* God, then there is only one God, no matter how many different religions exist, each with its own God.

We could only guess some answers to all our questions by observing the laws of nature. These laws apply to all objects and all the life forms on this planet. The closest comparison to our civilization would be a model, based on some living bacteria culture. What we could learn from it is not encouraging - beginning is always followed by its destructive ending.

Appendix A   The Universal Time

In 1913 *Willem de Sitter* proved that the light is propagating through the universe with its constant speed, relative to the stationary vacuum filling the universe, and is independent of the speed of its light source [1].

That excluded the effect of *aether wind* [2]. The speed of light was also excluded from the suspected variables. Since the speed of the orbiting Earth is also constant, the only culprit left was the time.

The concept of time was always an enigmatic subject and at the time of the experiment, science accepted the concept of time put forward by *Isaac Newton*, named after him the *Newtonian time* (described already in previous chapters). This concept defines the time flowing in one direction only and at constant, unchangeable rate.

This rigid, unchangeable time could have not been used to explain the mysterious results of *Michelson-Morley* experiment [2], and *Albert Einstein* therefore discarded this concept of *Newtonian Time* entirely.

He used calculations derived from an abstract experiment put forward by *H. A. Lorentz*. [3] In its essence, this experiment was simple to follow, but its correct interpretation is more difficult.

The experiment consisted of a light beam, sent to a distant mirror, and then the time for the beam to return was calculated.

Lorentz found that there was a substantial difference between time of arrival t_0 of the beam sent from a stationary light source, and the time t_1 of the beam sent from a moving source.

He then calculated and defined their ratio, t_1 / t_0 as *Lorentz Factor* Y:

$$\gamma = \frac{1}{\sqrt{1 - \dfrac{v^2}{c^2}}}$$

c is the speed of light in vacuum
v is the speed of the light source

It seems to me that for many of us it is difficult to distinguish between the delay in time, what *Lorentz Factor* represents, and slowing the rate of time flow. It is obviously clear that the *Lorentz Factor* does not slow, or speedup the rate of time flow at all.

This formula is incorrectly used to calculate the slowing of the rate of time flow on a fast-moving object. For example, it was believed that during interplanetary voyages, the time in a moving spaceship will proceed at a slower rate than the time on the Earth. For some hard to understand reasons, *Albert Einstein* also used *Lorentz Factor* Y to define the concept of relativistic mass m_r:

$$m_r = Ym_0 = \frac{m_0}{\sqrt{1 - \dfrac{v^2}{c^2}}}$$

[1] explained in *appendix B*
[2] explained in *appendix D*
[3] explained in *appendix D*

With increasing speed of the light source, the relativistic mass also increases and should the speed of the light source increase to c, then the relativistic mass becomes undefined. All that only because the light, when compared to a stationary observer, will reach a moving observer in delayed time?

This simple formula was also wrongly interpreted that "*nothing could move faster than the light because its mass becomes infinitely large*".

The concept of relativist mass cannot be correct and any calculations involving this concept cannot be correct either. This is particularly true for the mathematical derivation of Einstein's famous equation $E = mc^2$ [1].

We can conclude:

- In our universe reigns the *Universal Time*, i.e., *Newtonian Time*, which does not change its rate of flow. The only time which changes its rate of flow is the *Subjective Time,* as described in previous chapters.

Appendix B   The Double-star System

In 1908, *Walther Ritz* suggested that the light progresses through the space with constant speed c, relative to its source. That was refuted in 1913 by *Willem de Sitter*, who based his conclusions on observations of a double-star system.

He reasoned that if the speed of light c was relative only to its source, then if observed from different parts of the orbital path, the light from the star would travel away and toward us at different speeds.

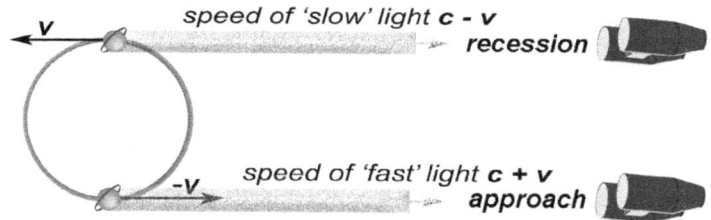

Fig. B1 *W. de Sitter - double star system.*
During approach, the revolving star moves toward the observer, and during recession it moves away with speed v.

If the light emitted by the orbiting star changes its speed at which approaches the *observer* on our planet, then the *observer* would see that '*the "fast" light given off during approach would overtake "slow" light, emitted during a recessional part of the star's orbit.*'[2]

Since this is not the case, *Willem de Sitter's* observations imply that the light in vacuum must be propagating with its constant speed c, regardless of the speed of its source.

[1] explained in *appendix E*
[2] From Wikipedia, *W. de Sitter* - Double-star System.

This plausible conclusion was soon incorrectly interpreted as '*nothing can move faster than the light*' and '*no matter how fast the observer travels, the light will be always passing with its constant speed*'.

Albert Einstein also characterized these observations in his book as: '*VII THE APPARENT INCOMPATIBILITY OF THE LAW OF PROPAGATION OF LIGHT WITH THE PRINCIPLE OF RELATIVITY... By means of similar considerations based on observations of double stars, the Dutch astronomer De Sitter was also able to show that the velocity of propagation of light cannot depend on the velocity of motion of the body emitting the light. The assumption that this velocity of propagation is dependent on the direction "in space" is in itself improbable.*' [1] The *special theory of relativity* has been with us for almost a hundred years, and it seems that it is infallible. However, in an essence its credibility relies heavily on:

1. the failure of the *Michelson–Morley experiment*,
2. the results of *Willem de Sitter's* observations of a double-star system,
3. the results of *Lorentz's* calculations.

Walther Ritz's observations eliminated the speed of the light source from factors affecting the speed of the propagation of the light through the vacuum. The already known fact that in different materials the light propagates with different speeds, decisively points to the content of the vacuum, as the factor limiting the speed of light.

It is very revealing to compare the vacuum, filling the universe, to different restrictive media, especially if we add the sound to the picture.

One of the remarkable differences between the sound waves and the light waves is that with increased density of the restrictive medium, sound speeds up, but the light slows down.

	Sound	Light
Vacuum	-----	300,000,000
Air	343	300,000,000
Water	1,490	225,000,000
Glass	5,600	200,000,000

Since the vacuum limits the speed of light, it must have some speed-limiting properties, as for example the water and glass have. This support the presence of the *aether* in the vacuum, which is still debated topics.

Accepting the conclusions drawn from *Walther Ritz's* observations, we could state that the light propagates through the universe with its constant speed $\underline{c}$, which is independent of the speed of its source.

[1] Book 'RELATIVITY THE SPECIAL AND GENERAL THEORY'
    by *Albert Einstein*, 1920

Appendix C   Lorentz's Transformations
"Reality continues to ruin my life."
Bill Watterson

Well before *Albert Einstein* defined his *special theory of relativity*[1], Dutch physicist *H. A. Lorentz* was already attracted by the relationship between the light and time. He conducted an abstract experiment, in which he used the light, propagating relatively to a universal reference frame with constant speed c, to calculate a delay in time caused by the observer's movement.

In the first part of his abstract experiment in the *universal reference frame*, i.e., relatively to the stationary medium in which the light propagates, a stationary *observer* sends a beam of light over distance S_0 to a distant mirror, and measures the time t_0 it takes for the beam to return.

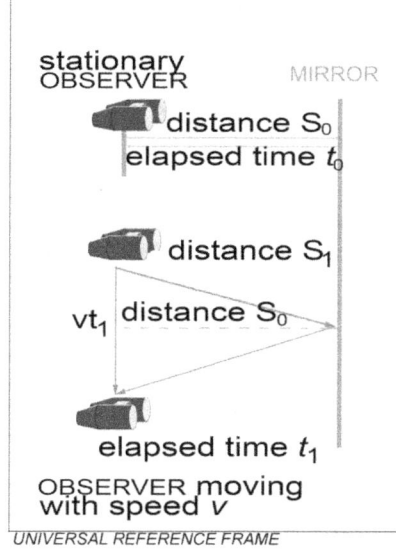

Fig. C1 *Lorentz abstract experiment, with a stationary and moving observer.*

In the second part of this experiment, an *observer* moves on a straight line with speed v, sends a beam of light to the mirror and measures the time t_1 it takes for the light to return. The light travels the distance S_1 which is greater than S_0. The resulting *Lorentz Factor Y* then describes how much longer it takes for the light to reach a moving *observer,* instead of a stationary *observer.*

[1] Book 'RELATIVITY THE SPECIAL AND GENERAL THEORY' by *Albert Einstein*, 1920, Ph.D., translated by *Robert W. Lawson*, M.Sc. University of Sheffield,

Calculations of *Lorentz Factor.*

$$t_0 = \frac{S_0}{c} \quad t_1 = \frac{S_1}{c} \quad \text{and} \quad S_0^2 + v^2 \cdot t_1^2 = S_1^2 \quad \text{therefore} \quad S_1 = \sqrt{S_0^2 + v^2 t_1^2}$$

$$t_1 = \frac{\sqrt{S_0^2 + v^2 t_1^2}}{c}$$

$$t_1^2 = \frac{S_0^2 + v^2 t_1^2}{c^2}$$

$$t_1^2 \cdot c^2 = S_0^2 + v^2 t_1^2$$

$$t_1^2 \cdot c^2 - v^2 t_1^2 = S_0^2$$

$$t_1^2 (c^2 - v^2) = S_0^2$$

$$t_1^2 = \frac{S_0^2}{(c^2 - v^2)} = \frac{t_0^2 c^2}{c^2 \left(1 - \frac{v^2}{c^2}\right)} = \frac{t_0^2}{1 - \frac{v^2}{c^2}}$$

$$\frac{t_1}{t_0} = \frac{1}{\sqrt{1 - \frac{v^2}{c^2}}}$$

v/c	Lorentz Factor γ		v/c	Lorentz Factor γ
0.5	1.15		0.99999	223
0.8	1.7		0.9999999	2236
0.95	3.2		0.99999999	7071
0.98	5.02		0.999999999	22360
0.99	7.08		1.0	? (infinity)

Fig. C2 *Some selected values of the Lorentz Factor.*

Lorentz in his calculations assumed that relative to the *universal reference frame*, the speed of light is constant. He also assumed that the *observer's* clock, measuring the time delay, is not affected by the *observer's* speed. (This possibility was mistakenly introduced by the *special theory of relativity.*)

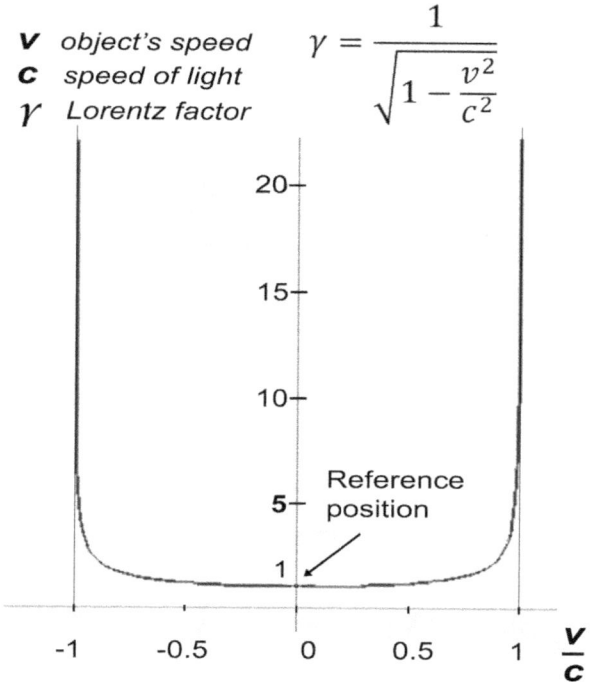

$$\gamma = \frac{1}{\sqrt{1 - \dfrac{v^2}{c^2}}}$$

V object's speed
C speed of light
γ Lorentz factor

Reference position

$\frac{V}{C}$

Fig. C3 *Graph of Lorentz Factor.*
v *is observers' speed and* **c** *is the constant speed of light.*

It is important to note that all that *Lorentz* achieved with his calculations was to calculate the time delay only. For the light wave, progressing with constant speed, he calculated the time difference between reaching a moving *observer* instead of a stationary observer.

It is also important to note that in this experiment, the flow of time does not change. During the *observer's* travel, the same clock with the same rate of time flow is used, as when the *observer* is stationary. Should the time flow on this clock change, as is generally believed, then *Lorentz's* calculations would be meaningless.

In his calculations, *Lorentz* also does not specify if the clock was stationary or travels with the *observer. For Lorentz* this would not make any difference.

In the following years, the first incorrect assumption made by many physicists was to mistake this delay in time for a change in the rate of time flow. They believed that the rate of time flow, measured by the *observer*, observing a beam of light, would change due to the *observer's* movement.

The second incorrect assumption was to consider the *Lorentz Factor* in only one special case: The *observer* is moving along a straight line, perpendicular to the line connecting the *observer* and the light source, and from the position closest to the light source.

That excludes any other movements, but in reality, the *observer* could move in whatever direction, and from whatever position.

The simplest case to investigate is the *observer's* movement along the line, connecting the *observer* with the light source. The movement could be in both directions, toward and away from the light source. This option traveling on a collision line, directly toward, or away from the light source, was missing entirely from *Lorentz's* experiment. This situation is illustrated in the following figure:

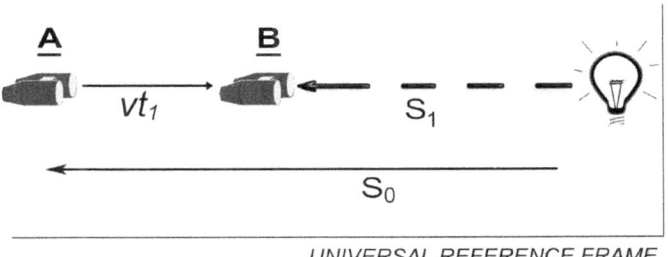

UNIVERSAL REFERENCE FRAME

Fig. C4 *Observer traveling on a collision line with the light source.*

Observer moves from **A** to **B** in time t_1. The light travels distance S_1, and the *observer* vt_1.

The light emitted by the bulb will reach the stationary *observer* in time t_0 and cover distance S_0. To reach a moving *observer*, the light would have to cover distance S_1 in time t_1, while the *observer* would travel the distance vt_1

1. Stationary *observer*:
   The light will reach the *observer* at position   **A** in time t_0
   Distance traveled by light will be $S_0 = ct_0$
2. Moving *observer*: Moving with speed v from position **A** to **B**.
   At **B** the light will reach *observer* in time t_1
   Distance traveled by the light $S_1 = ct_1 = S_0 - vt_1$
   In time t_1 the *observer* will move $AB = vt_1$

Again, the similar calculations could be used as used by *Lorentz*:

$$t_0 = \frac{S_0}{c} \qquad t_1 = \frac{S_1}{c} = \frac{S_0 - vt_1}{c}$$

$$t_1 c = S_0 - vt_1$$
$$t_1 c + vt_1 = S_0$$

$$t_1 = \frac{S_0}{c + v}$$

$$\frac{t_1}{t_0} = \frac{\frac{S_0}{c+v}}{\frac{S_0}{c}} = \frac{c}{c+v} = \frac{1}{1 + \frac{v}{c}}$$

Now we have two formulas, both depicting the *Lorentz Factor.*
Original *Lorentz Factor:*   Extended *Lorentz Factor:*

$$\gamma = \frac{1}{\sqrt{1 - \frac{v^2}{c^2}}} \qquad\qquad \gamma' = \frac{1}{1 + \frac{v}{c}}$$

In the graph below, the right-hand side of the dotted line represents a situation, where the observer moves on a collision course with the light source. In this scenario the *Lorentz Factor* **Y'** will infinitely decrease. That means the light will reach the observer in a shorter period of time than in a situation, where the observer is not moving directly toward the light source.

Full line represents the original formula for calculating the *Lorentz Factor* **Y**. Dotted line represents the extended formula of the *Lorentz Factor* **Y'**.

This dotted right-hand part of the graph is vastly different from other parts and yet, all that makes such a difference is only a very slight change in the direction the observer travels. The second formula includes not only delay, but also reduction in time interval, needed for the light to reach a moving observer. The difference between these two factors is obvious.

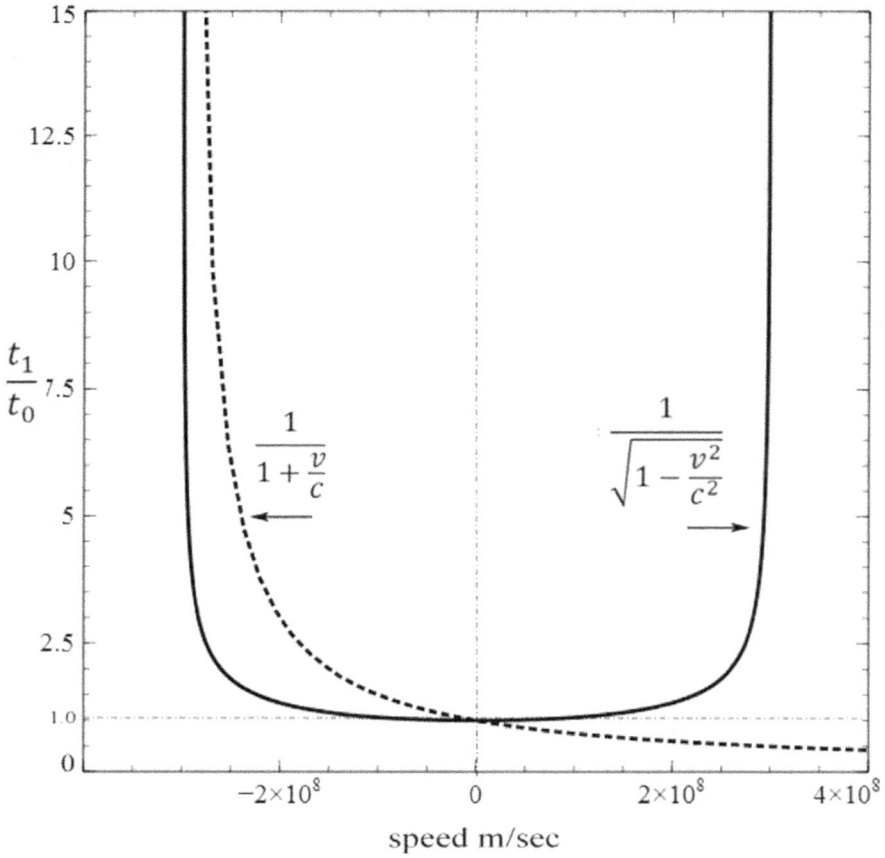

Fig. C5 *Two versions of Lorentz Factor.*

Should we incorrectly use this extended factor to calculate the rate of time flow, then the time will speed up, which is contrary to what was deduced in the *special theory of relativity*. This factor is valid only for observers moving towards the mirror on a direct line, connecting both the observer and the light source.
It is also easy to prove that the *Lorentz Factor* will change with the starting position, as illustrated in the following figure.

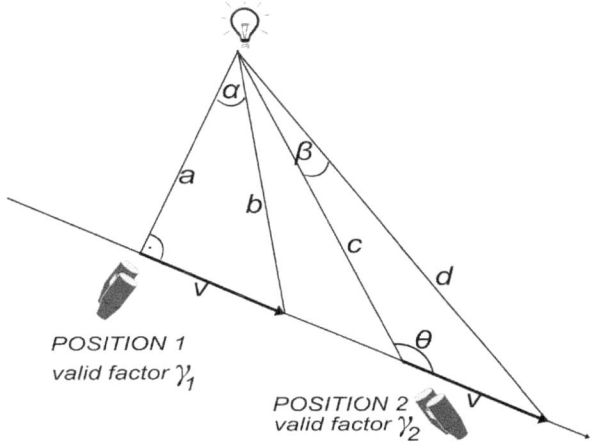

Fig. C6 *Observer moves with speed v in the same direction.*
Separate measurements are taken for position 1 and 2.

Observer moves with speed **v** in the same direction. Separate measurements are taken for position **1** and **2**. Using some properties of a triangle and some trigonometric functions, we could compare these two ratios.

$$\frac{v}{\sin \alpha} = \frac{b}{\sin 90} = b \qquad \frac{v}{\sin \beta} = \frac{d}{\sin \theta}$$

$$\sin(180 - \theta) = \frac{a}{c} = \sin \theta$$

For both factors to be equal: $\dfrac{b}{a} = \dfrac{d}{c}$

$$\frac{\frac{v}{\sin \alpha}}{a} = \frac{\frac{v \sin \theta}{\sin \beta}}{c}$$

$$\frac{1}{a \cdot \sin \alpha} = \frac{\sin \theta}{c \cdot \sin \beta}$$

$$\frac{1}{a \frac{v}{b}} = \frac{\sin \theta}{c \cdot \sin \beta}$$

$$\frac{b}{av} = \frac{\frac{a}{c}}{c \cdot \sin \beta} = \frac{a}{c^2 \sin \beta}$$

$$\frac{b}{a} \neq \frac{av}{c^2 \sin \beta}$$

49

Starting from position **1**, comparing **_b_** and **_a_** will produce the value of the originally defined *Lorentz Factor Y*.

Should the *Lorentz Factor* describe movement initiated at any other position on that line, for example position **2**, then the ratio of **_d_** and **_c_** should be the same as **_b_** and **_a_**.

We can choose any position and evidently, the angle **_β_** and resulting distance **_c_** could have any value, provided **_α_** > **_β_**. As a consequence of that, the value of **_a/b_**, which proportionally represents *Lorentz Factor Y*, could vary with the position, and can have an infinite number of values. The correct formula for the *Lorentz Factor Y*, would be also different and would have to include the angles **_β_** and **_θ_**.

The following example illustrates the general case, when the observer could move in any direction, not just on the line perpendicular to the line connecting the observer and the light source.

The observer could move with the same speed **_v_** to any of the positions **_P1_**, **_P2_** and **_P3_**. To reach the *observer*, the light has to travel a different distance, and therefore it will reach the *observer* with different delays.

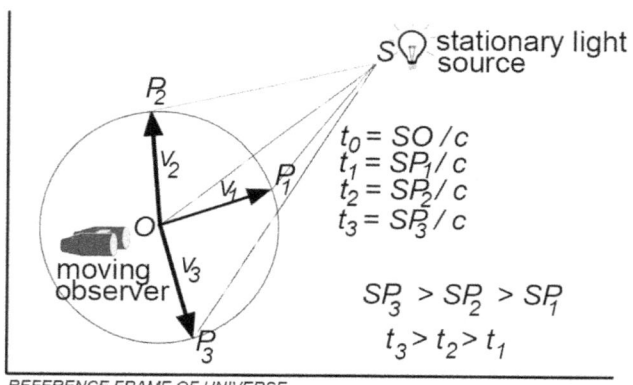

$$t_0 = SO\,/\,c$$
$$t_1 = SP_1\,/\,c$$
$$t_2 = SP_2\,/\,c$$
$$t_3 = SP_3\,/\,c$$

$$SP_3 > SP_2 > SP_1$$
$$t_3 > t_2 > t_1$$

REFERENCE FRAME OF UNIVERSE

Fig. C6 *Observer could move with the same speed* **_v_**
to any position **_P1_**, **_P2_** *or* **_P3_**

Starting from position **_O_** and moving to positions **_P1_**, **_P2_** and **_P3_**, the distances traveled: $OP_1 = OP_2 = OP_3$

When the *observer* is stationary, the light will travel the distance
$$OS \text{ in time } t_0 = SO/c$$

Similarly: $t_1 = SP_1/c$ $t_2 = SP_2/c$ $t_3 = SP_3/c$

Since $SP_1 < SP_2 < SP_3$, the light will travel a shorter time interval, therefore:
$$t_1 < t_2 < t_3$$

The *Lorentz Factor* is defined as a ratio of time taken by the light to reach a moving *observer*, to time taken to reach a stationary *observer*.

Then, for different directions of travel and the same *observer*'s speed, we would have different values of *Lorentz Factor:*

$$Y_1 = (t_1\,/\,t_0) \qquad Y_2 = (t_2\,/\,t_0) \qquad Y_3 = (t_3\,/\,t_0)$$
resulting in $Y_1 < Y_2 < Y_3$

These differences are not due to the different observer's speed.
Since $\underline{v}_1 = \underline{v}_2 = \underline{v}_3$, they would have to be calculated using a different formula for Lorentz Factor **Y**.
We have already calculated one such factor **y'**, for a simplified situation, and obviously, the difference is substantial.
It is obvious that the values of the *Lorentz Factor* depend not just on the *observer's* speed $\underline{v}$, but also on the position and direction the *observer* is heading. The starting position and direction of movement plays a vital role, and if the *Lorentz Factor* should be used in any calculations, it has to be included in the formula.
The following is the simplified diagram of the *Lorentz* experiment. Observer travels in two different directions and for each situation, different versions of the *Lorentz Factor* exist.

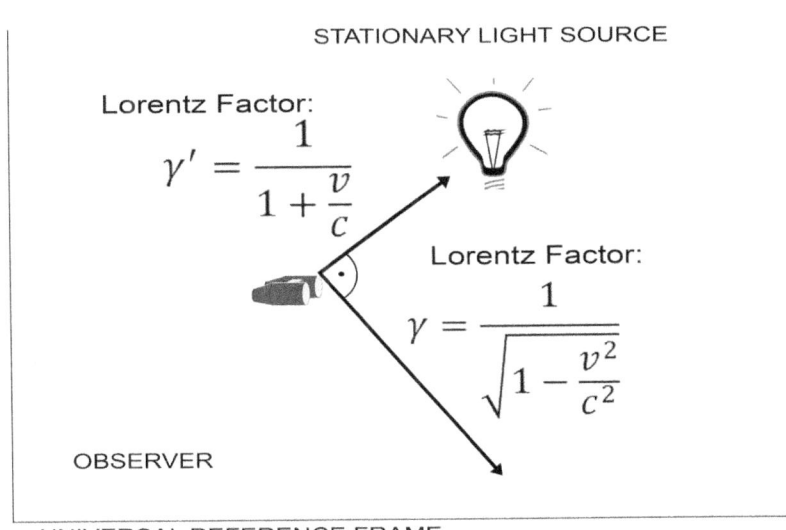

Fig. C7 *Simplified diagram of Lorentz experiment.*
Observer travels in two different directions and for each situation, different versions of the Lorentz factor exist.

We can conclude:

- *The Lorentz Factor does not represent any changes in the rate of flow of time. Furthermore, it is incomplete and has no use in any real-world calculations.*
- *To use the Lorentz Factor to define the relativistic mass is erroneous.*

Appendix D   Michelson-Morley experiment

"I always tried to turn every disaster
into an opportunity." John D. Rockefeller

Modern science is very uneasy with any mysteries and does not wish to accommodate their existence. Yet, one mystery is still with us and what's more, it was created by science itself!

Towards the end of the 19th century, the scientific community was divided in two camps: One believing that the universe is filled with a substance called *aether* and the opposing camp believed the universe is filled with nothing. There was no proof supporting either belief and finally, in 1887, two English scientists *Michelson* and *Morley* decided to eliminate this uncertainty, once and for all.

Fig. D1 *Michelson's interferometer.*

They were encouraged by the latest discoveries involving the light, and decided to use it to confirm the presence or the absence of the *aether* in the universe.

They used an apparatus with two arms, forming a right angle between them. At their intersection was a prism, separating the beam of generated light in two beams and sending each beam to the mirrors fixed at the end of each arm. The expectations were that the returning beams will form a light pattern on the display, observable through the eyepiece.

During the experiment, the arms were slowly rotated, as illustrated on the following diagram:

STATIONARY AETHER

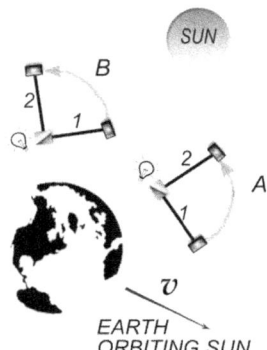

Fig. D2 *The rotating interferometer stationed on Earth.*

The general consensus was that ideally, both light beams arrive at the eyepiece at the same time. In the presence of the *aether*, the longitudinal beam sent in the direction of the moving Earth will be exposed to an '*aether wind*', and it will encounter greater *aether* resistance, and should therefore arrive at the display slightly later than the transversal beam.

In the figure *D2* are illustrated two positions, *A* and *B* of the instrument. In the position *A*, the arm *1* is pointing longitudinally in approximate direction of the orbiting Earth, i.e., against the apparent *aether wind*. The created resistance will slow down the progressing light on arm *1* more than on arm *2*. It was expected that in the first position *A*, the beam on arm *1* should arrive at the display slightly later than the beam on arm *2*.

This discrepancy in the time of arrival of both beams would then affect the light pattern, formed on the display and observed through the eyepiece. To ascertain that the arm *1* points in the same direction as the moving Earth, the arms of the instrument were rotated, while the displayed light patterns were observed.

At some point during the rotation, arm *1* was definitely pointing in the same direction as the moving Earth, and arm *2* was pointing directly to the Sun. At that point, the possible resistance of the *aether wind* would be greatest. That should be seen on the reflected light pattern as a distinct interference of two waves, arriving at slightly different times.

The instrument was turned even further to position *B*, and after every changed position the resulting observed interference should have been obvious. At least that was generally believed and expected. To a great surprise, no interference was observed and the conclusion was that both light waves arrived, more or less, at the same time:

'But when the experiment was made, it was found that the two beams arrived back at the same time. ... Now it must be recognized at once that this was a most extraordinary thing. Here was an experiment, performed with every care and apparently with full understanding of what was being done, which completely failed to give the result that common sense would have thought inevitable.

... If any explanation is to be given, therefore, it must necessarily involve something revolutionary.' [1]

Evidently, something 'extraordinary' and 'revolutionary' always deserves a sensible explanation. Since the whole situation is part of our physical world, all relevant physical laws must apply. Then the convincing justification, in some understandable and logical terms, for the obtained results has to be found.

At the time of the experiment, *Michelson, Morley* and all the others in the scientific community, based their conclusions on a simple, common sense example, where a body traveling against the wind will be slowed down by the wind more than the body traveling in a transverse direction.

That, for example, would be true for the sound traveling through the air, due to the physical nature of the sound waves and the propagating medium. For the same reasons probably, it was wrongly assumed that:

- The light progressing in longitudinal direction, i.e., against the *aether wind*, would take longer to return to the eyepiece than in the transversal direction.
- Without the *aether wind* , both beams arrive at the eyepiece at the same time.

[1] Described by *Herbert Dingle* in 1922, in his book 'Relativity for All'.

Based on the apparent failure of this test, the conclusion was that there is only a vacuum filling the universe and no *aether*. This conclusion is still considered as valid even now, and the presence of the *aether* in the universe is not generally accepted.

At the time of the experiment, it was believed that the speed of light will be affected by the friction created by light moving through the *aether*, i.e., the speed of light will slow down.

It was also assumed that the Earth and the instrument are both moving with the same speed $\underline{v}$, relative to the universal reference frame, and the Earth's orbiting speed $\underline{v}$, which is approx. 30 Km/sec, is used in calculations.

Corrections to original calculations were later made by *Alfed Potier* and *Hendrik Lorentz*, which proved that both beams do not arrive at the eyepiece at the same time, as was originally presumed. The corrected calculations proved that even in the absence of the *aether*, for the constant speed of light $\underline{c}$, the longitudinal beam always arrives later than the transversal. The whole experiment and all the corrected relevant calculations could be presented in a simplified, corrected scenario as illustrated by the following figure.

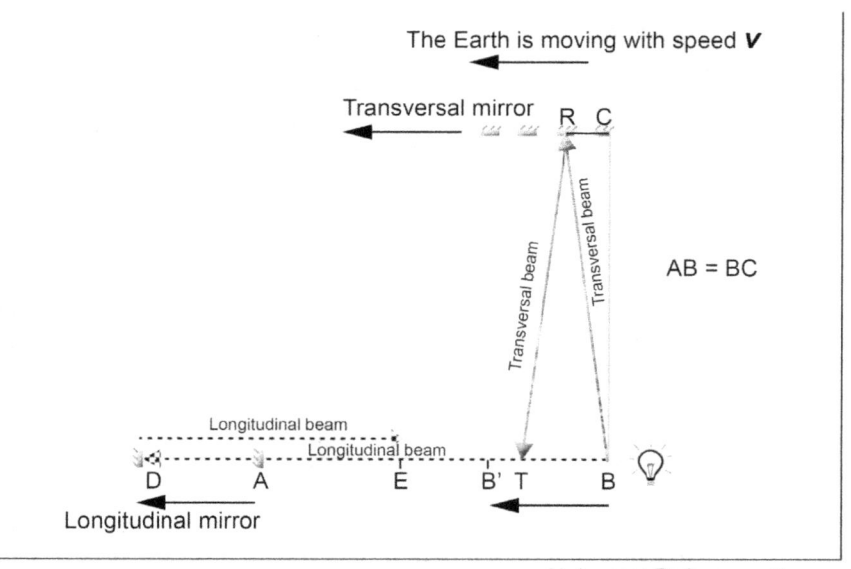

*Fig. **D3** Corrections done to Michelson-Morley experiment.*
The longitudinal beam of light from the source will travel
*from **B** to **D**, and then reflects to **E**.*
*Before reflecting, the starting position **B** moves left by **AD**, to position **B'**, and*
*after reflecting moves to position **E**.*
*The transversal beam will also start at **B**, travel to **R** and then reflect to **T**.*
*Both mirrors, the light source and the eyepiece will move with speed **v**.*

In figure **D3** the traversal beam is sent to traversal mirror in time t_1 at speed c covers distance $ct_1 = BR$ => $t_1 = BR/c$

Since $BC = AB$

$$BR = RT = \sqrt{RC^2 + BC^2} = \sqrt{RC^2 + AB^2}$$

$$t_1 = \frac{\sqrt{RC^2 + AB^2}}{c} \quad => \quad t_1^2 = \frac{RC^2 + AB^2}{c^2}$$

Since $RC = vt_1$

$$t_1^2 = \frac{(vt_1)^2 + AB^2}{c^2}$$

$$t_1^2 c^2 - v^2 t_1^2 = AB^2$$

$$t_1^2 = \frac{AB^2}{c^2 - v^2} \quad => \quad t_1 = \frac{AB}{\sqrt{c^2 - v^2}}$$

Total transversal time $t_T = 2\, t_1$ => $t_T = \dfrac{2AB}{\sqrt{c^2 - v^2}}$

Time for the longitudinal beam to travel consists of two time intervals t_1 and t_2.

$$t_1 = \frac{BD}{c} = \frac{AB + vt_1}{c} \quad => \quad ct_1 = AB + vt_1 \quad => \quad t_1(c - v) = AB$$

$$t_1 = \frac{AB}{c - v}$$

$$t_2 = \frac{DE}{c} = \frac{AB - EB'}{c} = \frac{AB - vt_2}{c} \quad => \quad ct_2 = AB - vt_2$$

$$t_2 = \frac{AB}{c + v}$$

Total longitudinal time $t_L = t_1 + t_2$

$$t_L = \frac{AB}{c - v} + \frac{AB}{c + v} = \frac{AB(c + v) + AB(c - v)}{(c + v)(c - v)}$$

$$t_L = \frac{ABc + ABv + ABc - ABv}{c^2 - cv + vc - v^2} = \frac{2ABc}{c^2 - v^2}$$

In the graph in figure **D4** the distance **AB** was set to 9 m[1], since that was the original length the light traveled.

The full line represents the longitudinal movement, with corresponding values of t_1, the time of arrival of the longitudinal light beam. The dotted line represents the transverse movement, with corresponding values of t_2, i.e., the time of arrival of the transverse light beam.

Both precise moments of arrival of each light beam depend on the speed v, at which the source of the light, stationed on the Earth, is moving through space. For low speed v there will be only a negligible difference between t_1 and t_2, and only at higher speeds the difference becomes noticeable.

Since the speed of orbiting Earth's is 30,000 m/sec, the time and distance traveled by:

Longitudinal beam   time taken   6.0000035266687399e-8 sec
                     distance traveled 18.000010580006219 m
Transverse beam   time taken   6.0000017633341110e-8 sec
                     distance traveled 18.000005290002331

Difference   time   1.7633346289845457e-14
                     distance   ...

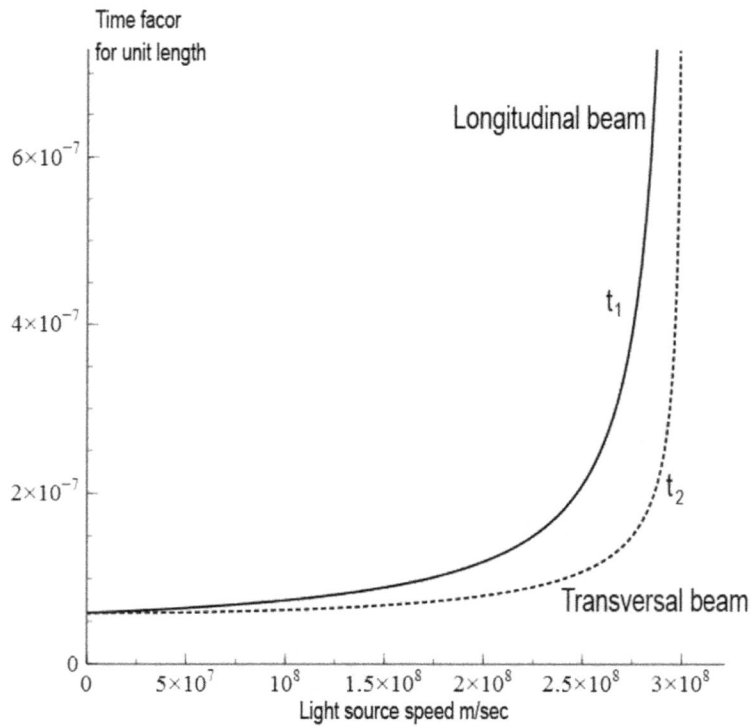

Fig. D4 Michelson-Morley experiment.
For a specific Earth's speed of 30,000 m/sec, the time taken by the longitudinal beam to return is longer than for the transverse beam.

[1] For more precise results, in the experiment the beam was reflected across the arms more than once. For a greater simplification, we used 9 m length only.

It is obvious that even without any interference from the *aether wind,* it takes longer for the beam of light to return from longitudinal direction than from transversal direction, i.e., the beam from transverse direction arrives sooner than the beam from longitudinal direction.

This result, produced one year later by corrections done to the initial calculations by *Alfed Potier* and *Hendrik Lorentz,* supports the original idea that two waves arriving at different times should produce interference patterns on the display. Yet, the results of observations did not support these expectations, since during the experiment there was no interference observed.

Michelson expected that Earth's motion would produce a pronounced shift in light pattern, but in the published results of this experiment, the greatest separation they achieved was only 0.018 fringes.

In the experiment, the speed of orbiting Earth $\underline{v}$ was considered to be 30 000 m/sec., but in reality this speed does not represent the real speed of the Earth in space. It is believed that the speed of the solar system in the universe is 200 Km/sec, which was not included in the original calculations, since such an increase would have an insignificant impact on the calculated results.

For the apparatus arms' length of 9 m, the difference between distances travelled by transversal and longitudinal beams represents approx. 10 of a typical light wavelength of light, i.e.,500×10^{-9} m. For greater accuracy, *Michelson-Morley* later reflected each beam many times before combining them on the eyepiece. That would make the difference in the distance travelled by each beam grater and it would easily accommodate some additional beams sent by the light source.

. Shall we change the length of both arms to, for example, 90 m, we will obtain a slightly different results:

Longitudinal beam	time taken	6.0000060000060005e-7
	distance traveled	180.00018000018002m
Transverse beam	time taken	6.0000030000022500e-7 sec
	distance traveled	180.00009000006750 m
Difference	time	3.0000037505329433e-13 sec
	distance	9.0000112521693154e-5 m

Now the difference between distances travelled by transversal and longitudinal beams represents approx. 180. typical light wavelengths of light.

What that means - before the arrival of the longitudinal beam, additional transversal beams arrive at the eyepiece. The displayed pattern, if any, would then not correspond to the pattern created by two beams, originating at the same time from the same light source and arriving at slightly different times.

This corresponds to what *Morley and Morley* saw during the experiment - the eyepiece was flooded by many returning transverse beams, producing almost a uniform light.

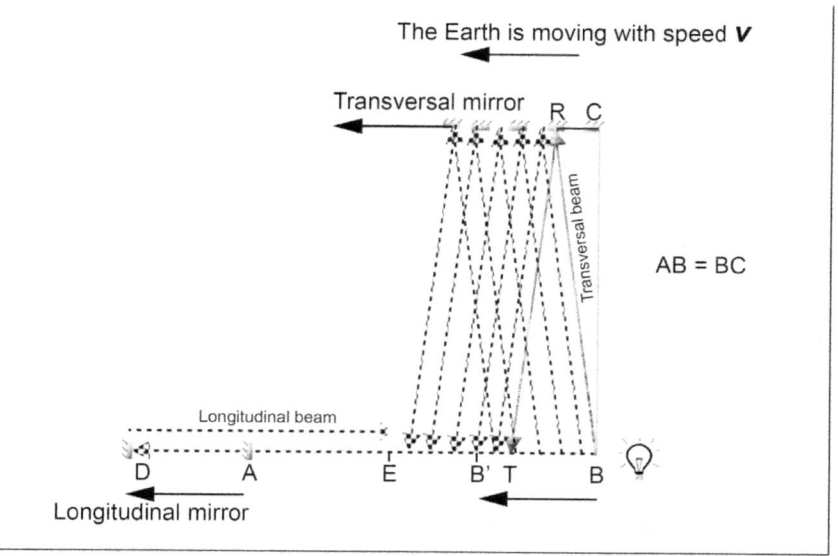

Fig. D5 *Michelson-Morley experiment.*
Before the longitudinal beam arrives at the eyepiece, additional transversal beams could arrive at the eyepiece.

To sum up this analysis:

At the time of the experiment, the explanation of the unexpected results was that the transverse and longitudinal beams arrived more or less at the same time and therefore the space is not filled by the *aether*. This possibility was disputed later by the corrected calculations, which clearly confirmed that the difference in arrival time of beams is real and valid.

Although this is a serious flaw in the design of the experiment, the main problem was that the beam of light was treated as a one-dimensional entity, adequate for some theoretical calculations. Since one-dimensional entities do not exist in our world, therefore this approach is not applicable for real situations.

In our physical world the source of light does not consist of one singular point, sending only a single, one-dimensional beam. It consists of many of such points, each sending a unique light beam towards the mirrors. These point sources are spread over the whole three-dimensional area of the light source and therefore the created beams are not synchronized and the distance they travel also slightly differs.

It was therefore wrongly assumed that only one transverse beam of light reflected from the transverse mirror will reach the *observer* at point *F*.

In reality, the *observer* could see some other beams, reflected from the transverse mirror, together with the returning longitudinal beam.

This explanation agrees with the results achieved by the *Michelson-Morley* experiment and explains its apparent failure. Unfortunately, the failure to achieve the expected results was at that time interpreted as non-existence of the *aether*, which by itself could not be considered as a valid explanation.

For the experiment's apparently peculiar results, some other explanation had to be found:

'Various suggestions were offered, but, in the light of future investigations at any rate, none of them was so satisfactory or far-reaching as the most revolutionary of all—the principle of relativity.' [1]

In this experiment three basic attributes were analyzed:

- the light and its speed,
- the speed of moving apparatus,
- and the flow of Universal Time.

By consensus it was incorrectly agreed that the culprit must be the time. *Albert Einstein* then included this conclusion in his *special theory of relativity*. He wrote about this experiment: *'If the Michelson–Morley experiment had not brought us into serious embarrassment, no one would have regarded the relativity theory as a (halfway) redemption. '*

However, the failure of this experiment cannot be explained by slowing down the rate of flow of *Universal Time*, which is not possible. Furthermore, it proves nothing about the *aether*, which was the intended aim of the experiment.

The results of this experiment do not eliminate the existence of the *aether*, but they also do not prove that it exists. Despite all these shortcomings, the *Michelson–Morley* experiment is still one of the fundamentals of the *special theory of relativity*.

[1] Taken from *'Relativity for All'* by *Herbert Dingle*.

"Maybe in our next lives ..."
My grandmother Františka

The Predicament Song

We live in dimensions, three and the time,
We measure our success by mountains of gold.
We segregate ourselves into who's strong or frail,
Who's young and beautiful and who's ugly and old.

We became servile, content and weak,
Our freedom is dying by thousands of cuts.
We have no spine to hold our head high,
We have no courage, we have no guts.

We look for the features, so easy to see.
We don't see beauty, hidden in soul.
We value only the obvious traits,
We wrongly judge people, simply as all.

We introduced music, with no melodies,
We introduced art, for those who can't paint.
We regard artists as demigods,
We proclaimed a sinner being a saint.

To sustain our living we fabricate life,
We muster poor creatures through the devil's farm gate.
They never see sunshine, they never feel rain,
What they all live for, is a meal on our plate.

We're heading for our destruction, like a runaway train,
Without reverse and without brakes.
No moral restraints, we don't look back,
Just always forward, whatever it takes.

Our future is written all around us,
As ancient worlds all turned to dust.
We live like culture producing poison,
But we don't believe it, and in lies we trust!

Notes: